TOWER DOG

DOG

LIFE INSIDE THE DEADLIEST JOB IN AMERICA

DOUGLAS SCOTT DELANEY

Soft Skull

New Yor

Library of Congress Cataloging-in-Publication Data Is Available.

Cover design by Faceout Studio
Interior design by Tabitha Lahr

ISBN 978-1-61902-938-5

Soft Skull Press
1140 Broadway Suite 704
New York, NY 10001
www.softskull.com

Printed in the United States of America
Distributed by Publishers Group West

10 9 8 7 6 5 4 3 2 1

For Ali, who made anything possible.
For Brian, so some day he might understand.

"It isn't crab fishermen. It isn't ice road truckers or Northwest loggers or bayou coonasses wrestling alligators. It isn't cops or firemen or combat soldiers. It isn't oil field roustabouts or trapeze artists. According to OSHA and the body count—it's us."
—BRODY KILFOYLE, CEO of K.M.C.A.

"The fucking stupid son-of-a-bitch killed himself. Simple as that."
—ECGBERHT SCHMIDT, Owner of Sunburst Tower Systems

"Tower climbing remains the most dangerous job in America."
—OSHA ARCHIVE, United States Department of Labor

"Johnny was a brave young man. And I . . . for me . . . I need the world to know that for a cell phone call a brave young man has died. For high-definition television, very brave young men are dying. There needs to be some new regulations or laws. Maybe just basic companies caring. There needs to be some accountability."
—TED AND MAUREEN McWILLIAMS, Cumming, Iowa

AUTHOR'S NOTE

I began typing this story on January 4, 2013. Since then, thirty-five of my subjects have died, averaging one every 40 days.

And not from natural causes.

This story is true. Some names have been changed to protect the moronically guilty, the chronically modest, and the otherwise disengaged. Some locations have been changed as well.

CONTENTS

PROLOGUE
THE GREEKS HAD A WORD FOR IT

Kansas used to be an ocean. Stick a shovel deep enough in any part of the state and you are likely to unearth some plankton-sucking platywhoozitz big as Wall Street. On the weekends you can find geology and paleontology students lined up along the jagged cuts in the blue highways digging out the fossilized flora and fauna from eighty-five million years ago. Rudists, crinoids, squid, ammonites, sharks, bony fish, turtles, plesiosaurs, pteranodons, and—be still, my beating heart—giant clams. When the ocean receded it left behind a topography no one ever associates with Kansas, for Kansas is many countries. It is Ozark Country, high desert, expansive prairie, Monument Valley, Yorkshire Dales, wildflower heaven, and, in places, even deeply forested. People think Kansas is flat and unpleasant, and that is fine with Kansans because as Doc Wimmer, my biology teacher at Southwestern College, would say, *If Kansas still had an ocean, everyone would live here.* Part of that residual topography is the continent's largest remaining tract of tallgrass prairie, four million acres of it, that lies in the eastern third of the state. These are the Flint Hills.

The Flint Hills stretch almost border to border from Marshall County up by Nebraska to Cowley County in the south before extending down into Oklahoma. A portion of I-35 slices through the heart and most scenic section of the Flint Hills for 117 miles. The first time I ever drove through the Flint Hills, I had to pull over and stop to admire the austere beauty of it all. My typing cannot do the place justice. William Least Heat-Moon devoted 624 pages to the hills in his *New York Times* bestseller *PrairyErth*, so I am not even going to try to match words with that guy. But for those with an affinity toward the unforeseen density of empty places, the Flint Hills are the place to be, even though if you took your kids there on vacation they would seek to have you committed or at least sanctioned by the Ministry of Fun.

Each April during the burning season, the fires can be seen from space, and if you are fortunate enough to view this from ground level, you would get a new appreciation of hell on earth. This burning, which for tens of thousands of years was a natural occurrence, has only changed in its method of ignition. It used to be lightning (and still can be), but in our intrinsic human quest to control everything, the burning is now carefully orchestrated between the farmers and ranchers and several bureaus of land management, the fires started with various homemade and store-bought contraptions. But the end result is the same. Dazzling.

It took a while for the early homesteaders of the mid-1800s to appreciate this. Shortly after they slapped together sod houses along one of the few flowing creeks and began to turn the land, representatives of the Kiowa, Osage, or Kansa dropped by to tell them, *You might want to rethink this one.* The settlers did not take this in the spirit of high-plains fellowship it was meant, but (the white man's burden lying heavily on their shoulders) instead as being warned off the land by a bunch of Christless savages, a matter they would have to attend to as soon as the crops were in and they could scrape up enough wood to build a church. What the Kiowa, Osage, or Kansa and their progenitors had known for about twelve thousand years was that the creeks would dry up for months come the summer, the land was not very good for raising crops, and when the spring conflagration came, the settlers would be hauling ass back to the Missouri River landings up in St. Jo faster than they could say *land grant*. After they figured out that the fires were not being caused by hostiles, the settlers settled and each year tried to raise the same unraisable crops and went nearly insane with thirst and dug holes in the earth to escape the yearly inferno. Any wood for churches was soon resigned to keeping the hearths crackling throughout the heartless plains winters. But they stayed anyway.

The Greeks had a word for that. They called it *akrasia*. Loosely translated it means *the tendency to gravitate toward danger when there are safer alternatives available.* Socrates (in Plato's *Protagoras*) asks precisely how this is possible because if one judges action A to be the best course of action, why would one do anything other than A? I asked this question of myself on my very first tower just north of El Dorado, Kansas, in the winter of 1997. I looked up at the 240-foot guyed-wire tower and thought, *Why would one?*

"Second thoughts?" Angelo Kilfoyle said.

Angelo Kilfoyle, a.k.a. Power, due to what we yelled every time we needed some real heft behind a chore, had the teeniest *I know something you don't know* smile on his face when he said that.

"No shame in it," he said. "Just the dog."

At the time I lived about ninety minutes from that spot, so getting the dog[1] did not concern me at all. Getting up that tower did. I did not want to do it. I had been trained days earlier at my house, hanging from the pecan trees in my belt as Angie and Hangman familiarized me with the equipment. I had no problem with that; they told me I was a natural. They held back the fact that twenty-foot pecan trees shared little in common with 240-foot guyed-wire

---

1. *GETTING THE DOG:* Many times new hires are flown into market, but if they screw up, they might "get the dog," a long bus ride home.

towers. They knew, though they admired my self-deceiving confidence, that I would have to figure that out for myself. I had met Angie in a bar in Astoria, Queens, shooting pool, and for reasons we still cannot agree upon he threatened to kick my ass. Not in eight-ball but to *kick your igrant Yankee ass right here and now*. He was fifteen. I was twenty-seven. Power was what my dad would call a fireplug. Five-foot-nine-inches tall and stout from neck to ankles, biceps like hams and thighs like bigger hams. He had straight brown hair down to his ass that he put in a ponytail and, when it reached his butt-crack, he would cut it off and donate it for wigs for children undergoing chemotherapy. And he had a crystalline outlook on the world as a whole. Life was two things: fucked-up or hilarious.

The weather that had landed on us that day in El Dorado, Kansas, was fucked-up. My first climb was hilarious. I started out well, but by about forty feet up I was having second thoughts as I broke the tree line and the wind cut through me like the icy breath of death. It was twenty-two degrees on the ground, and I, wearing long johns, jeans, insulated coveralls, Thinsulate gloves, two pairs of wool socks, a hooded sweatshirt, and a woolen cap, was quite comfortable down *there*. This would not be so bad. But for the next two hundred feet and the next eleven hours it was very, very bad.

"Hangman's waiting on you," Power said, and he was, having glided up the tower a few minutes earlier without a care in the world.

The El Dorado tower was situated at the southern end of the Flint Hills just a mile east of the first of two I-35 exits. Directly to the north, El Dorado Lake stretched for seven miles and across it blew a forty-five-mile-per-hour wind straight into my face, pushing me back off the tower. According to windchill calculations it was negative eight degrees, and I wondered how in the world did Hangman get up this pig and if he was even still alive. Depressingly enough, that would not be the coldest I ever was on a tower. But, as Power would later say, "It's good when you think your first is the worst." The wind was blowing so hard that I could not hear the deafening blast of the horns from a northbound train until it was miles past the tower, the wind pushing the train clamor straight back into Oklahoma. And the wind itself was deafening, an unrelenting roar that after a few hours would make you go a little crazy. At one hundred feet it got even worse as dense wet fog enshrouded the tower, wrapping me in mist that soon turned to ice, as did the steel. And it was then and there I realized just what kind of insane bullshit I had gotten myself into. I could not see above me. I could not see below. I was colder than I had ever been in my life, and my muscles, all my muscles, my goddamn ear muscles were throbbing. I did not know if I could climb another ten feet let alone 140. And the wind. That bastard wind let me know there was nothing between me and Canada but hope.

And I never felt more alone.

Power must have sensed this. He got on the radio.

"You okay?" he said.

I tried to sound okay, but I wanted the hell off even though Christmas was a week away and I needed the money. No amount of money could be worth this shit.

"I can't see you," I said.

"You *can't?*" he said, chuckling like a patient mother smiling down at a five-year-old trying to tie his shoes. Then he said, "That's 'cause you're in a cloud, asshole."

Somehow, and to this day I do not know how, I made it to the crow's nest,[2] where Hangman was lying back in his belt, back to the wind, smoking a cigarette.

"You gotta smoke three to smoke one," he said. Actually neither of us *said* anything. We screamed it at the top of our lungs and still could barely hear each other. And now that I was up there, it dawned on me that I had no idea what I was supposed to do.

"Where do you need me?" I said.

---

2. *CROW'S NEST:* A structure in the upper part of the main mast of a ship. The term is also used in reference to topmost structures on towers. This crow's nest was a 12 x 12 x 12–foot triangle of steel with top and bottom railings.

"Just chill," he said, not meaning to be funny as he unhooked his lanyard and climbed over to where I was huddled against the steel, stiff as a brick and hurting.

"Don't *DO* that!" I said.

Hangman just smiled and belted off by my side. He pried my hands from the steel. "That death grip[3] will wear you out sure as shit," he said. "Relax. Trust your gear."

"Like you just did you fucking idiot!"

Another smile.

"I'm me and you're you," he said. "Relax."

I tried to relax, but I was a fast learner. I *learned* I wanted to just fucking die and *learned* this was the best I was going to feel for the rest of the day.

"Check this out," Hangman said.

He climbed outside the crow's nest, hooked his positioning lanyard around the steel at chest level, and let go with his hands.

"C'mon," I said. "Knock it off."

Then he let go with his feet and was punched off the tower and into the wind. It was me who had the two-second heart attack.[4] Hangman was laughing. He put out his arms like Superman and hung there, windsurfing, his toes pointed straight south.

He was flying.

"You try it," he said. After I explained to him he was out of his fucking mind, we went to work, though much like my climb I could not tell you exactly what we did that day except for that Hangman did most of it. I do remember that when I got back to sweet Mother Earth, the ground betrayed me, shifting beneath my feet, and I wobbled and fell on my side. Power picked me up and brushed the hard dirt off my face.

"You got to get your sea legs," he said. "That'll come."

---

3. *DEATH GRIP:* Common to new climbers, it is the tendency to grip the steel very tightly, and one is unaware one is expending a lot of energy and arm strength.
4. *THE TWO-SECOND HEART ATTACK:* What happens when a tower dog's belt, or lanyard, slips but an inch. At elevation that inch feels like a mile.

When I got back to the motel that night in El Dorado, I could barely move. I had played football in college not far from there, and I felt as if I just went sixty solid minutes on the losing side. I lay on the bed and stared at the ceiling and mourned myself, and then Power came in with a six-pack of longneck Budweiser. When he sat on the edge of the bed, it hurt.

"If you just lay there and stretch out every muscle you got over and over again, it won't hurt as much tomorrow. Trust me, it *will* hurt, but it won't hurt as much." Hangman came in a few minutes later.

"I didn't get much done, did I?" I said.

"Fuck that," said Angie. "You got up and you stayed up. When you stopped for that bit I thought for sure you were coming back down. But you didn't. So fuck you and welcome to cellular hell."

"Is it always like this?" I asked.

Power and Hangman shared a laugh.

"Shit NO!" said Hangman.

"Hell," said Power, "This is the worst fucking site we've ever been on."

From then on I was hooked. I had hung on top with one of the best of them and would live to tell about it, over and over again. My first tower. I pass it every now and then in my travels, and I always tip my hat and smile. If I did that, fuck it, I can do anything.

JULY 9, 2013

The Bismarck, North Dakota–area Occupational Safety and Health Administration office is investigating the deaths of two tower technicians who fell from a cell tower in Mountrail County. According to the Mountrail County Sheriff's Department, the workers fell 250 feet yesterday.

The agency said that the cell phone tower technicians were employed by Monarch Towers, Inc. Monarch's president, Jack Boone, was reportedly on his way to the accident site to assist in the investigation.

The company is based in Sarasota, Florida, where Boone also owns Broadcast Technologies, Inc. The men were members of a crew that was retrofitting the structure. The sheriff's department could not immediately provide information regarding the tower's owner.

Yesterday's fatalities brought 2013's total deaths of workers falling from communication structures to six. The last fatalities occurred on May 28, when two workers were killed in Georgetown, Mississippi.[5]

---

5. *Wireless Estimator,* "Two Techs Die in North Dakota Cell Tower incident," July 9, 2013. wirelessestimator.com/content/articles/?pagename=Wireless-Construction-News-7.13.

CHAPTER ONE MISSISSIPPI
DEATH IS A SALESMAN

I promised them death.

I sat in the requisite springy leather-and-chrome chair across from two programming executives from NBC network television, and I confidently, unabashedly, promised them death. The gulf between us was bridged by a svelte glass-topped coffee table, two feet wide, that might as well have spanned a light year. For the past four days I had been camped out in a borrowed condo at the foot of those crappy hills out by Azusa honing my pitch. I had the prose. I had the graphics. I had the video. I had the stats. I had the fucking sizzle. But I was worried. I had been in L.A. for a week, and this was the last of nine meetings. A late-afternoon meeting at the studio. At four thirty. On a Friday. In the world of that world, it could not have been a worse time to propose *any* project. The meeting had already been postponed three times, and by three thirty that afternoon in October of 2007, I suddenly ceased to care.

I walked through the studio lot, which was absolutely soulless, canyons of sound stages and empty lanes shouldered by dozens of charging golf carts. Canyons of suspect encouragement. The character actor Richard Herd once told me, "Kid [yes, he said *kid*], this town will encourage you to death." And fuck if he wasn't right. My usually unreasonable optimism was usurped by persistent whispers of futility. The frigging hangover I was sponsoring didn't help much. There were manatees break-dancing behind my eyes. On top of an overstuffed trash can by the closed commissary I saw a little yellow rubber duck bath toy. It was filthy. I put it in my pocket, smoked one last cigarette, popped in the gum, and made my way to my meeting. And I did not give one damn. I wanted to go home. Back to Kansas. Back to my little leaning farmhouse and my porch and the cool hummocks and the cotton-ball sky. I was already a failure. An also-ran. A writer of fiction and drama and film that fell the perfect victim to early success and subsequent mediocrity. "Writing is a lot like golf," an English teacher, Troy Boucher, once said to me. "You make one good shot and you think you're Arnold fucking Palmer."

I had a play in the Samuel French catalog that some high school would perform once a year and I would get a check for twenty-three dollars. Yes, it did play in Peoria. I had short stories in the lits, out of print and out of anybody's right mind. I had made a decent buck for a few years writing for every major studio and some smaller independents. Eleven screenplays that never got the green light. I was also a fixer, working on other writers' screenplays that never got the green light. In thirteen years I had one film make it to the screen. Limited release. So what the fuck. This was it. I was tired of it. I was sick of the finely dusted L.A. everywhere. October never looked so bland. I had been playing the game for thirteen straight years, and the box score said not only had I lost but I had lost resoundingly. I didn't give a damn, but, oddly, they know and respect when you don't give a damn.

After the intros I bounced in my chair and they offered me the bottled water and I asked for coffee, which was damn foolish at a studio at that hour. "How about tea?" one of them said. The execs were two bright women, one dirty-blonde and one brunette, capable and stunning women in their mid-thirties, I guessed, but in L.A. they could have been fifty and it wouldn't have mattered. They were ready to leave before I got there. Their shoulder bags were packed and waiting at their feet. I was looking for the one last big deal of my life. They were looking for something to fill one prime-time reality hour for the fall of 2008. They leaned in. Elbows on thighs. Hands clasped. They cocked heads and smiled expectantly. The brunette said, "*Sooooo?*"

And I had nothing. Absolutely nothing to say. I had said what I planned to say so many times that it had no jazz anymore. It was rote. I was rote. I was pathetic. Once more into the breach with my hat in my hand and my guns half-cocked. I dropped glossy copies of a nine-page proposal on the glass-topped table. They didn't want to read. They never want to read. They wanted to hear and, most importantly, to see. I knew that. "You can look at the hard copy later," I said. I opened my laptop, aimed it toward them, and selected the sizzle that Dennis Fallon and I had made back in Kansas City—the icon labeled *TOWER DOGS*.[6] I hit play and went to the bathroom.

"Don't Fear the Reaper" by Blue Öyster Cult is a track that lasts five minutes and one second. That was the song for my sizzle reel, and as I opened the bathroom door, I heard,

> *All our times have come*
> *Here but now they're gone*
> *Seasons don't fear the reaper*
> *Nor do the wind the sun and the rain*

I locked the bathroom door, turned on both taps, and flushed the toilet to mask the sound of me throwing up. I splashed my face with cold water. It wasn't nerves. I had been to too many meetings

---

6. *TOWER DOG:* Any one of the eighty-five hundred to ten thousand workers across America who earn their living on cell phone and broadcast towers, more formally known as *tower hands, tower techs, top hands, communications specialists,* etc. In some parts of the Northwest they are called *tower rats.* They are also known unilaterally as crazy bastards.

to be nervous. It wasn't even the beer from last night. It was the Maker's Mark that Fallon kept in the condo. The return volley of the Maker's from my gut to the toilet at NBC reminded me that I was not a whiskey drinker like a mother's slap to the back of the head. I took the dirty duck from my pocket and tried to wash off the black smudges, but it didn't work. I speed chewed four sticks of Big Red and returned to the execs. Blue Öyster Cult was at 4:16.

She had taken his hand
She had become like they are
Come on, baby
Don't fear the reaper

They did not look up at me. At 4:55 into the reel, my closer, my killer graphics and project tag, scrolled up the screen—a cell phone tower with Harley Davidson wings radiating microwaves like the old RKO masthead and the words:

TOWER DOGS
YOU'RE DYING TO TALK...AND IT'S KILLING US.

Silence. As if choreographed, both execs leaned back and crossed their legs. They looked at me now, two identical expressions, stony, one half of Mount Rushmore. It felt like I had somehow invaded. Like I had asked for their bra size. The dirty-blonde's face said, *Annnnnnd?*

I had a one-sheet. *THE* one-sheet to end all one-sheets. I handed it to them, saying, "Crab fishermen aren't drowning. Semis aren't crashing through the ice. Loggers aren't sawing through their femoral arteries." My one-sheet, a straight download from the Occupational Safety and Health Administration (OSHA) and the U.S. Department of Labor stating unequivocally that cell phone tower workers perform what is indeed the deadliest job in America, landed on the glass with the sound of thunder. The stats were undeniable. "If you follow these guys for one year," I said, "you can count on a fatality in the industry every 33.18 days. You can set your goddamn watch by it." And then these words came out of my mouth, these words that have haunted me since my vocal cords oscillated them into my oral history.

"People *will* die," I said.

The brunette's lips tightened, and her eyes slimmed. She stared at the dirty-blonde, whose hand had come up over her mouth and nose, not in shock but contemplation. Four eyes clicked on mine at once, and the brunette said, as if she were ordering a donut, "Can you bring this in for six hundred thousand dollars an episode?"

The entire exchange lasted seventeen minutes, and I sleep-walked back into the concrete canyons, which were now glistening cathedrals, with the biggest deal of my thus-far-so-called career. I had finally arrived . . . I was a hack. A soon-to-be wealthy hack. The empty lanes were paved with reality gold. I squeezed the dirty duck in my pocket. Fallon and I drove down to Venice. We sat under a bistro umbrella on the concrete boardwalk, he with his Maker's, me chain smoking and downing Coronas with lemon, not lime. I bought my dirty duck a shot of Cuervo. It was about forty-five minutes after I left NBC that my phone rang. It was Marc Shmuger, president of Universal Pictures, the man who had given me my first writing assignment in the movies, a man who had risen in a few short years from producer at Columbia's *Art of War* to one of the five most powerful men in Hollywood, a man who I unashamedly would visit on every pitch trip I made to L.A. looking for crumbs. Marc had no creative control over the decision made by NBC Network TV, but he is the one who set up the sit-down.

"Heard you hit a home run," he said.

When I hung up the phone, I grinned at Fallon. He knew instantly and ordered another round. I don't remember much of the rest of the night. But before we drank ourselves stupid, I stretched my legs down the walk for a few blocks. The big, smog-sodden sun was gliding down behind the Santa Monica Pier when it hit me—not only was I a hack, I was officially a whore. Coming from the world of fiction and drama, taught by serious men of purpose, if I was going to write for television I wanted to be in the class of Paddy Chayefsky or Paul Attanasio or Aaron Sorkin or David Chase or Rod Serling. But this was not going to be that kind of work and I did not have that kind of chops. And as I meandered back to Fallon through the henna-tattoo stands and blankets covered with gourd bongs a line from *Speed-the-Plow* softly reminded me that *you're a fucken' bought-and-paid-for whore, and you think you're a ballerina.*

I promised them death.

I promised them death, but nothing could have prepared me for just how spectacularly I would make good on that promise.

AUGUST 13, 2013

An unconscionable tenth death this year occurred last evening when a Georgia man fell approximately two hundred feet to his death off of a monopole near Coats, North Carolina. It appears that he was not 100 percent tied off.

Officials say John Dailey, 49, of Silvester lost his balance while trying to hook off to a tower member and fell. EMS personnel determined he was deceased on the scene. The body was transported to Betsy Johnson Memorial Hospital.

Dailey and another tower technician were working on the structure at approximately 5:20 P.M. when the accident occurred. The Transmit crew was performing installation services for Sprint under the direction of Alcatel-Lucent.

The tiger team was changing out a radio when Dailey fell. According to workers who knew Dailey, the tech had approximately twenty-five years of industry experience.

The 225-foot monopole, constructed in 1999, is owned by SBA Communications. A spate of fatalities has hit the industry in past weeks. Six technicians have died since July 8, 2013.[7]

---

7. *Wireless Estimator,* "Tenth Tech Succumbs Following 200-foot NC Fall," August 13, 2013. wirelessestimator.com/content/articles/?pagename=Wireless-Construction-News-2.14.

CHAPTER TWO MISSISSIPPI
JUNGLE BOY SAYS BELT UP

I never called myself a writer. I never introduced myself as a writer, and I was always uncomfortable when I *was* introduced as a writer. During certain heady intervals in my life I have indeed made money as a writer, but I never came close to making a living at it. Actually paying for the stuff of life. My father was one of the very first Green Berets, but you would not know that to talk to him. He was basically a trained assassin, but when you asked him what he did in the Green Berets, he would say, "I jumped out of planes." I did not even know he was a Green Beret until I was eleven and one day found an assortment of bayonets and Ka-Bars in an old metal milk box in the attic. He used them to stir paint. He did not define himself as that Green Beret. He defined himself as a construction worker for Otis Elevator.

I would remember him when out on an L.A. pitch trip at a Starbucks virtually surrounded by a convention of writers pecking away on their laptops, their overlapping cacophony of crap regarding their pitch or their project or their development deal too much to endure. Writers, actors, singers—my own sadly expectant subspecies. You've seen us. We show up on *Jeopardy* like bad coins. I am a Groucho

Marxist—declining any membership in any club that might have me. I never wanted to be at the loaf-heel end of the old joke

What do you do?
I'm an actor.
Really? What restaurant do you work at?

My father had the most finely calibrated bullshit detectors of any man I ever met. His litmus test for discerning simple basic honesty in people was asking one question: *Where are you from?* His theory was that if you ask someone where they are from, the first answer will almost always be a lie, albeit a white one. This was particularly true of New Yorkers (where he was from), Angelinos, Chicagoans, and Texans.

Where are you from?
New York.
Where in New York?
The City.

BUZZZZ! In his mind nobody was ever *from* New York City. They all moved there. He was from Brooklyn but would never say he was from *The City.* Two questions later, the truth would hobble out.

Where in The City?
Brooklyn.
What section?
Canarsie.

So fucking say you are from Canarsie, for Christ's sake. Dad's theory included the hypothesis that 90 percent of the time the person was truly *from* at least thirty miles from the nearest big city. I asked him why people did that, and he said people just like to attach themselves to something bigger than they are. Human nature. They want to be from someplace better. The City beats Canarsie any day. They want to be *something* better. "Be from where you are from," he said. "And be what you become." I put Dad's hypothesis to the test all the time. Mostly on airplanes.

Where are you from?
L.A.
Where in L.A.?
The City.
Where in The Ci—?
Azusa.

At Starbucks I could be downright cruel.

What do you do?
I am a writer.
Really? What have you written?
I just finished a screenplay.

Then the wings come off the fly.

Wow. Who for?

So when people asked me what I did, I would say, "I'm a tower dog." And that is exactly what I was on July 9, 2007. That morning, I sat on the back porch of my rented ninety-year-old farmhouse about seven miles southeast of Lawrence, Kansas. Fallon and I had just finished postproduction of *All Roads Lead Home,* the only film I ever had made. Limited release. He had begun shopping it around the festival circuit, and I was flat broke. Meagan was inside tending to our seven-month-old son, and she knew I had to find work and fast. My choices were limited.

I stared at the miles of unbroken country to the west, south, and east. Windbreaks of hedge and cottonwood, acres of fallow pasture, ancient and rusted farm machinery half-buried in the land, all wrapped about a ten-acre watershed. The vista was little different from when the house was built, with one exception—a thousand-foot guyed-wire communications tower[8] about five miles northeast as the crow flies. Even in the blaring late-morning sun, the blinking white beacons on the tower were clearly visible. At dusk the photocell would kick in, and the red lights would blink until dawn. This was to keep the local livestock from going nuts at night, the chickens from careening off their laying cycles.

It was 102 degrees, and I did not want to be outside. But inside my house, I, a man who had installed hundreds of cellular phone systems on towers in thirty-seven states, could not get a cell phone signal. Ironic? Yes. Job security? Absolutely. Outside was not much better, but if I held the phone just right . . .

At 4:00 P.M. I finally found a sweet spot, and I called Brody Kilfoyle, president and CEO of K.M.C.A. (Kilfoyle Microwave & Cellular Associates) in Nashville. I had not worked for Brody in almost three years, but we spoke often. We were friends long before he started up his first tower crew back in 1997. Then he had a white Ford cargo van, a ten-foot box trailer, and a crew consisting of him, his little brother Angie, a house painter/musician/dancer named Clay, and Angie's friend from kindergarten—the Hangman. And three dogs. By that crackling hot afternoon in July of 2007, Brody had built a company boasting a fleet of trucks and trailers; a boom truck; a warehouse-and-office complex on Navaho Road in Nashville and one in Atlanta; over one hundred men in the field and twenty-four office employees, largely composed of friends and relatives; not to mention a small army of subcontractors plying the trade

---

8. *GUYED-WIRE TOWER:* A straight shot to the sky, usually consisting of identical twenty-foot steel sections bolted together and anchored to the earth by steel cables at three or more anchor points. These are the giants. The tallest guyed-wire tower in America is the 2,063 footer in Blanchard, North Dakota.

from Spearman, Texas, to Cape Cod. I came into the fold about six months after he started. I was the fifth Beatle. A woman I did not know answered the phone.

"I need to speak to Jungle Boy."

"Excuse me?" she said.

"I need to speak to Jungle Boy."

"And whom may I say is calling?"

"He'll know."

I heard the phone shuffle across her blouse and some muffled chatter.

"Are you sure you have the right number?"

"Yes. Who are you?"

"Lanai."

"Hi, Lanai—who are YOU related to?"

She'd had it.

"*EXCUSE* ME?!"

"Just tell Jungle Boy that Delaniac is not a writer again."

Long before the towers and the movies and NBC and the deaths, Brody and I had met in New York City. Yes—The City. Though I had moved to Kansas in 1989, I, like my father, was born in Brooklyn, and I grew up in Levittown, Long Island. I had met Brody by a dumpster in the freight elevator on the fourth floor of a building on West Twentieth Street between Fifth and Sixth Avenues in Manhattan, about half a block from the Limelight disco. I was a playwright at the Working Stage Theater Company. He, freshly arrived from Baton Rouge, was an actor taking soap opera classes from Julie Bovasso. We shared the same rented space. But what I *really* did was bust my ass sixty hours a week at John Grace, a pipe shop in Bethpage, and Brody was knocking out gypsy construction throughout the boroughs while moonlighting as a male stripper all over the Island, Connecticut, and Jersey. Brody lived out by the Elmhurst Tanks and, prone to locking himself out of his apartment, would simply and *somehow* scale the three-story building to his window and kick in the screen, much to the horror and consternation of his landlady. I once caught him swinging from a decrepit fire escape ladder trying to get a foothold close to his apartment window.

"Who are you?" I yelled up to him. "Fucking Tarzan?"

Hence the name Jungle Boy. He called me Delaniac for reasons I would rather not go into. Brody came on the line.

"How's Hollywood?" he said.

"I'm calling YOU, ain't I?"

Brody laughed with his whole body, and I could see him through the ether, baby blues sparkling and his shoulders silently bumping up and down.

"Copy that, brother," he said.

"What do you got?"

"Jersey."

"You kidding me?"

Brody and his minions were almost entirely Southerners and detested work above the Mason-Dixon Line. So I knew there had to be a lot of money in it.

"Man, we are on FIRE up there," he said. "You need?"

"I do need."

"Belt up," he said.

There was a small pause, and he knew what I was thinking but always had a hard time saying.

"This is the part where you ask me to advance you two weeks per diem," he said.

Within the hour I had packed my Explorer with three weeks' worth of clothes and my climbing harness. My tools. My dog tags, a clam knife, and my dad's Ka-Bar. I did not want to go. I had left the tower business five or six times swearing to never return. But it wasn't about me anymore. It was about the little seven-month-old baloney loaf Meagan was cradling in the kitchen. Meagan had that face, half smile half tears, the face only women can make that says *I don't want you to go but I know you have to and we'll be just fine don't you worry*. She knew enough not to say *be careful*. To tower dogs that was like saying *Macbeth* in the theater.

Brody wired me nine hundred bucks, and I picked it up at Western Union on my way out. I ran around Lawrence for a few hours, picking up this and that, the most important thing being a gallon of milk at the last QwikTrip on the outskirts of town. I poured out the milk and cut off the top two inches of spout with my dad's old knife. This was my piss jug. No self-respecting tower dog rolls without one. I shot east on K-10 as the sun set in my rearview mirror. Out the passenger-side window just west of Eudora, I watched as the white flashing lights on that thousand-foot guyed-wire tower clicked over to red. Over twelve hundred miles and twenty-three hours later, road-dopey with fatigue, I checked into the Extended Stay hotel on Rte. 17 in Ramsey, New Jersey, paid for one night only, hit the bed face first, and slept until 4:00 A.M. I had not eaten or slept in thirty-six hours, but when you mobilize that is precisely what tower dogs do.

It was still dark when I stepped into the hotel parking lot. I could tell which of Brody's employees were in market[9] just by looking at the trucks and equipment, which had taken over the place. It was ALL of them, judging by the rigs with Tennessee, Florida, Georgia, and Louisiana plates. I knew there were sub crews there as well—tags from Rhode Island and Massachusetts. They would muster at six, and I wasn't waiting around. I was starving.

I slid into the red Naugahyde booth at the Horizon Diner a few miles up Rte. 17 and ordered six eggs, bacon, sausage, ham, toast, coffee, orange juice, and water. I had not been in a real diner in years, and it was only then I grew nostalgic for the Northeast. Where else in the world could a person get kielbasa, souvlaki, half a smoked chicken, pecan pie, crab cakes, and a boiled lobster at four thirty in the morning? I spread out the *USA Today* and after skimming

---

9. *MARKET*: Any city or part of the country where upgrades or new installations are being performed. A tower dog can find himself "in market" for a few weeks or as long as several years.

the headlines went directly where I always went—the section where news from all fifty states is thumbnailed into two or three sentences each. Under the state of Kansas, it said:

KANSAS. JULY 10, 2007. TWO TOWER TECHNICIANS DIED AT 9:30 A.M. AFTER FALLING APPROXIMATELY FIVE HUNDRED FEET FROM A TOWER SOUTH OF KANSAS 10 NEAR EUDORA.

My fork froze over the ruptured yolks of my sunny-side-ups. When the waitress asked if she could freshen my coffee, I couldn't make words.

Mississippi OSHA Area Director Clyde Payne has confirmed that a tower technician fell to his death Saturday morning.

According to Payne, the 45-year-old man was a resident of Scott, Louisiana.

The technician fell approximately 125 feet at 8:30 A.M. OSHA began their investigation shortly thereafter.

The worker was employed by Custom Tower LLC of Scott, Louisiana. Owner Rick Guidry informed Wireless Estimator that he would not be providing any comments to the media.

The business was founded in 2006.

The accident occurred at 5181 Highway 149. It was not immediately known who the tower owner was and who Custom Tower was providing services for.

Payne said he was upset to learn of the fatality, the third climber to die in Mississippi this year. On May 28, 2013, two men were killed in Georgetown when gin pole rigging reportedly failed.

"We went a long time without any [deaths], and we did well for a good while. It's sad," said Payne.[10]

---

10. *Wireless Estimator,* "Weekend Accident Takes the Life of a Tenth Climber," August 19, 2013. wirelessestimator.com/content/articles/?pagename=Wireless-Construction-News-2.14#tenth.

CHAPTER THREE MISSISSIPPI
TERMINAL VELOCITY

Physics. It is an accepted rule of physics that shit will fall at 32.2 feet per second. Whether it be a bowling ball, a pogo stick, or a goddamned grand piano, the shit will fall at 32.2 feet per second. That is for the *first* second. After that, the rate of speed will increase an additional 32.2 feet per second for every second of free fall until maximum (or terminal) velocity is reached. Maximum or terminal velocity is reached at 177 feet per second. Physics. Inexorable and unforgiving.

The *USA Today* item was inaccurate in two ways. The tower was north of K-10, not south, and the two workers (I would find out later) fell from a height of seven hundred feet. The part about the workers being dead was bang on. I called Meagan, and she recited the full story printed in the *Lawrence Journal-World*. During their descent, Jerry Case, fifty-four, of Kansas City, and Kevin Keeling, thirty-three, of Independence, Missouri, did attain terminal velocity. And at nine thirty on the morning of July 10, as I was humming through the Ohio plains on I-76 out of Columbus at sixty-five miles per hour, Jerry Case and Kevin Keeling hit the earth at 120.69 miles per hour. They were the seventh and eighth industry fatalities so far that year, bringing the average to one dead tower dog every 23.87 days, far more drastic than the average I would later predict to NBC. There would be three more deaths by Christmas. The year before, nineteen workers had met such a fate, and as of this writing, since 2003 (and the keeping of reliable statistics), 130 tower dogs have plummeted to the earth from heights ranging from twenty to twelve hundred feet. They died in Utah, California, Louisiana, New Hampshire, Illinois, Arkansas, Michigan, Pennsylvania, Kentucky, Massachusetts, North Carolina, Tennessee, Indiana, Kansas, Washington, West Virginia, South Carolina, Georgia, Mississippi, Florida, Iowa, Wyoming, Virginia, Alabama, New Jersey, Oklahoma, New York, Arizona, Texas, North Dakota, Maryland, and Puerto Rico. And this was only in America.

I knew the work was dangerous. I knew we did the deadliest job in America and possibly the world, but it was something we never talked about. Once in a while someone would say, *Did you hear what happened in North Dakota?* But that was the extent of it. We didn't play it up, and the industry certainly did not mind that dim light being hidden under a bushel. Since 2005, the American public had been shown the deadliest jobs in America were ice road truckers and crab fishermen. Ice road truckers weren't even on OSHA's radar, and what the producers of *Deadliest Catch* never quite explained to its viewers was that though crab fishermen were indeed in the top ten deadliest jobs, those statistics included *ALL* fisherman, a fraternity of which crabbers are but a small segment. Misdirection aside, dead is dead, and being number one, Brody Kilfoyle would say, "is a distinction I could do without."

We never talked about the mortality rate because we knew it would never happen to us. *We* were trained. *We* were serious. Accumulatively we had tens of thousands of man-hours on thousands of towers, and the men under Brody's umbrella had not once claimed a serious injury. That is one hell of a safety record. Nor did we have any sympathy for those who *did* take the big drop because invariably it was their own damn fault. We called it *pilot error.*

> *Did you hear what happened in North Dakota?*
> Stupid bastards.
> *Guy in Texas fell 380 feet and hit the ice bridge. Cut him in half. Guts spilled over the generator.*
> What he have for lunch?
> *Dog fell a thousand feet in Iowa. Plowed three feet into a soybean field.*
> Shoulda' just buried him there and saved his family the heartache.

In fact, the majority of all tower fatalities happen because the worker either neglected his PPE[11] or forgot or ignored his training, largely the latter. If you follow your training it is *virtually impossible* to fall off a tower. A much smaller percentage of fatalities are due to equipment or structural failure. And THAT was our biggest fear. Brody, me, and all the employees of K.M.C.A. were certain we would not kill ourselves, but failures outside our control scared the shit out of us. We were confident, not cocky. And we were damn fucking good. *We* broke the rules all the time, but we were experienced. If some fly-by-night moron broke the rules and subsequently his neck, *fuck him, he deserved it.* Of course none of us ever *knew* anyone who had fallen. We couldn't put a face to the splatter.

As I slid out of the booth at the Horizon Diner, I folded up the *USA Today* and went to work. The parking lot at the Extended Stay was abuzz with tower dogs as I pulled in. Diesel trucks idled as job trailers, flatbeds, box trailers, and goosenecks were hitched. Crews were loading steel and spools of coax from the corner of the lot. Project managers were lining out construction managers who

---

11. *PPE:* personal protection equipment. An OSHA-approved variety of full-body climbing harnesses, lanyards, clothing, headgear, footwear, eye and hearing protection, etc.

were lining out crew leaders who were lining out top hands and groundhogs.[12] Men unloaded material from the two rental PODS they had dumped in the vacant lot as others humped their backpacks and lunchboxes and thermoses from room to truck cab. I sat in the Explorer just agape at the energy and extent of the activity. *Jesus Christ*, I thought. *"On fire?" Shit—this is a full-blown company mobilization.*

Not only had the boys completely taken over the parking lot and the vacant lot next door, but the tools of the trade were spilling into the ornamental shrubbery. The boom truck was parked there, and a backhoe on a flatbed. There was steel on sawhorses, a welding area, two propane-powered barbecue monster grills, and chairs and tables and pop-ups. A major contractor, GeoDyne Tech, had hired K.M.C.A. and a subsidiary of K.M.C.A., Sunburst Tower Systems, to basically wire all of New Jersey with the latest AT&T cellular upgrade. The boys were rolling in four- and five-man crews all over the state from Toms River to Moonachie, from the Amboys to Parsippany. As I walked into the melee, the first thing I heard was

"Fuck, Delaney, how the hell could you live in a shithole like this?"

It was Ansobert "Devil Anse" Schmidt, a wiry tower tech who I knew to be not only the stuff of legend but also a fucking lunatic.

"I live in Kansas."

"But you're FROM here, you damn Yankee."

If you stared straight down Rte. 17 toward NYC, thirty-one miles away, you could actually make out the Empire State Building. I pointed to it and said, "I am FROM the other side of THAT."

"Don't matter," Ansobert said. "It's *all* shit."

To Devil Anse, anything not within half a day's drive of McDonough, Georgia, was shit. He did not hate the North, *hate* being too forgiving a word; he *reviled* the North, where people couldn't do

---

12. *GROUNDHOG:* A tower dog who works primarily on the ground, usually too fat to climb anymore.

anything right. They didn't know how to put in a real day's work, didn't know how to drive, and certainly didn't know the first thing about iced tea. At five foot eight and 145 pounds, he was all sinew and snake-muscle, skin like salted leather, pound for pound one of the strongest men I have ever met. He once (after putting in fourteen hours on a three hundred–foot tower) climbed a two thousand–foot tower just to see how fast he could do it. The highest I have ever climbed is 680 feet, and it wasn't for shits and giggles. It about killed me. That two thousand footer that Devil Anse scaled was 546 feet taller than the Empire State Building.

His work ethic was an obsession: the run and gun, hit the lot screeching with the first glimpse of the sun, mobe[13] to the tower, head up and stay up till sunset, yelling the entire time. All day. Every day. His temper matched his ferocity for work, and he would fight at the drop of a hat, once conquering three opponents on the second-floor balcony of a Red Roof Inn in Sarasota, tossing one of the poor unsuspecting bastards over the rail. Disrespect him on the open highway and he would not think twice about taking his truck and trailer and pressing you right off the road.

Actually, I was surprised to see him there. Devil Anse didn't want to install antennae-line for GeoDyne Tech on existing towers—he wanted to *build* those towers. That was his forte. He and his older brother Ecgberht "Sarge" Schmidt were partners in Sunburst Tower Systems, along with Jimmy Tanner and Scotty Hamisch. When Brody closed a contract for the construction of a tower (or dozens of them), these were the guys who would build it. They were specialists, and Brody could always count on them to bring in a "stack"[14] faster than anyone in the industry. Those guys being here and not out stacking towers meant that *this* market was important.

Devil Anse barked at his crew, three guys I had never met, to get the show on the road. I looked at the new faces, and I was familiar

---

13. *MOBE:* To mobilize. Move. Get your shit and your men where they need to be as soon as possible.
14. *STACK:* The process of erecting a new tower.

with the expressions of dread. I knew what they were in for. Devil Anse would drive them into the ground that day, his one volume being screaming. All day. Every day. I waded into the quasi-organized chaos to get a handle on things.

I exchanged insults with the hands, some of whom I hadn't seen in ten years, and it was as if I saw them yesterday. Crack Baby was munching on the first of six candy bars he would eat that day. White Chocolate was shadow-kickboxing with an azalea bush. Preening himself in the side mirror of a Ford dually, Vic with a D was buttoning up one of his many collared NASCAR shirts. Bo and Gunn checked the rims of their Stetsons, holding them up to the sky. Bo once got us all kicked out of a motel in Spanish Fort, Alabama, because the maid moved his cowboy hat. He raised such hell the cops came. Super Mario and Bigfoot were humping channel steel to the horses. Ricky Boots was lacing up those knee-high leather boots. Two brothers, Mo and Ron, sat on the curb sipping Red Bull, the coffee of youth. But there were a lot of new faces among the forty or so workers, a situation that would come into play in a very bad way in the months to come.

Small talk. I didn't mention the USA Today item. Not something to talk about before work. That was a subject that, if it came up at all, would arise during the nightly "debriefings" in the parking lot. When the last of the trucks rolled, Ecgberht Schmidt, Jimmy Tanner, and Scotty Hamisch stood arms across chests discussing the day ahead. These were the bosses. These were the big men, in both position and stature. There was not a tower they could not build nor a piece of heavy equipment they could not run in their sleep. And indeed we had thrown many a wrench at Sarge from the tower as he sat in the running winch, smoked-out cigarette clamped in his teeth, hands folded atop his beer-keg belly, fast asleep. *That's* confidence. These were the men you wanted on the ground when you were hanging off a two-inch piece of steel at 280 feet in the air with a crane with three hundred feet of stick swinging seven thousand pounds of tower section within inches of your head. No place to run. No place to hide. Your life was literally in their hands and their hand signals. Their climbing days were long gone, though between the three of them, they had more air time, perhaps with the exception of Jungle Boy, than the rest of the company combined. Behind their backs we called them the *Three Wide Men*.

"What the fuck we supposed to do with you?" Sarge said, shaking my hand. "Fucking floater."[15]

His leg-of-lamb forearm had a tattoo of a self-support tower from his wrist to his elbow. Sarge, wearing his dirty blue T-shirt and stained khaki pants, broad face stubbled, looked like the guy at the end of the bar who delivered rocks all day and captained the bowling team at night. But he had a photographic memory and a fierce intellect. He could calculate the challenging mathematics of complex rigging, wind load, and flying steel in his head. He was a walking slide rule, and when in that mode his eyes would narrow to slits and his mouth would form silent words as he held up one hand, which encouraged all to *shut the fuck up*.

"I *could* sit around here with you fat bastards all day," I said.

They had rented an extra room and stuffed it with computers and printers so they could keep up with the phenomenal and sometimes ridiculous paperwork that went into a single job site. Scotty went back inside to attack the backlog. Sarge jumped in his truck and rolled. He would spend all day and hundreds of miles scooting from job site to job site, cracking the whip and surveying progress. That left me and Mr. James Tanner, a.k.a. the Godfather.

I had met Jimmy in the late '90s when Brody and his one crew were helping wire up the exploding Atlanta market with cellular service. He was walking through one of the many satellite shops we

---

15. *FLOATER:* A worker, not assigned to one crew, who replaces men who go on "break" about every six weeks. Though most tower dogs will take advantage of their company-paid-for six-week break, many have stayed out for years at a time. There are a pair of brothers who have been on the road for roughly fifteen years. When I started out in El Dorado, Kansas, I did so as a floater.

worked out of, and I noticed he was wearing a Kansas City Chiefs jacket. Jimmy didn't work for Brody at the time but for CES, another company using the same warehouse. We exchanged a few words, and I didn't see him again for almost six years. When we did meet up again, doing microwave hops across the Oklahoma panhandle, he remembered where I was from, where I lived, what job I was on in Atlanta (Downtown #1), my wife's name, and that I was a Chiefs fan. Whereas Sarge had a photographic memory for the work, he couldn't give a rat's ass about the details of my life. The Godfather retained such information. He could make even the lowliest tower dog feel important and necessary. They called him the Godfather because Jimmy had already put in thirty-five years in a fairly new industry and was still going strong. There was nothing he hadn't done. He was a walking encyclopedia for the uninitiated. When he wasn't running a boom truck or operating a crane or a winch or a Lowell, he was fielding calls from all over the nation, often from men he hadn't worked with in years, on exactly HOW to rig this or HOW to get that much steel up a mountain without a road or HOW to wire up the fourteen wires in a top-mounted beacon without electrocuting yourself. He didn't need photos. "Tell me what you're looking at," he would say. Then he would pause thoughtfully and say, "Now, what you *DON'T* want to do is . . ."

Jimmy, unlike many bosses in the industry who believed that tower dogs were *not paid to think*, encouraged thinking. When faced with a difficult problem on a tower you could get on the radio to Jimmy and the first thing he would say was, "Well, how do *YOU* think we should do this?" It didn't matter if you had five years in service or five weeks. On one of our many cross-country jaunts he said to me, "It never fails. I have been doing this for over thirty years, and every once in a while some slap-dick greenhorn will say, *Why don't we try it THIS way?* and I'll be damned if the little fucker wouldn't be right."

Jimmy was a teacher and a student of his vocation.

"We were not expecting you," Jimmy said.

That was typical Brody. Dispatch my ass across seven states in twenty-four hours and fail to let the bosses know I was coming. Jimmy had begun welding, fabricating a 10 x 4–foot steel mount (or *billboard*) for a tower where the specs were not quite what the blueprints said. Things like that did not slow him down. He had built, decommissioned, and retrofitted hundreds of towers, and little things like the structural prints and analyses being fakakta were not going to stop him from getting antennas in the air. He was as meticulous as always. He was the only person I ever met who could oil, cut, weld, grind, and polish steel without making a mess. It didn't matter that he was working in a vacant lot in Northern New Jersey. When he was done, you would never know he was ever there.

The Godfather was one anal motherfucker. And that discipline saturated his life. His paperwork was immaculate. When sitting down to eat, he would arrange his placemats and menu just so and then take the napkin and clean his own silverware. The dogs called him the Truck Nazi because he insisted on all company vehicles being cleaned and properly maintained at all times. This, of course, only happened when he was around, and it was a source of never-ending angst for him. His own truck looked like it just rolled off the factory floor, and he could not understand why the men simply could not live up to his expectations. Whereas most tower dogs' trucks and motel rooms and general appearance were succotash, Jimmy did not mix his peas with his carrots. Frankly, nobody ever lived up to Jim's expectations, and that was mainly because nobody could really do anything as well as he could.

Oddly, when we did fuck up, which was a daily occurrence, he didn't sweat the big shit. Once I changed a tire on a loaded job trailer and forgot to tighten down the lugs. It could have been a disaster, but he caught it and just said, "You forget something, Delaney?" Another time, in Fort Supply, Oklahoma, we were raising a ten-foot microwave dish in thirty-five-mile-per-hour winds. NOBODY in their right mind would do that, but it was a Sarge and Devil Anse job, and damn if they weren't going to get it done. The dish, 450 pounds of steel, was tagged off to the hitch of a pickup truck seventy-five yards from the base of the tower. The wind caught the dish, the tower groaned and lurched, and the pickup truck just bounced up and down and was dragged on its rear wheels for forty yards. Jimmy got on the radio and calmly stated, "Might want to rethink this one, fellas."

The angriest I have ever seen him, or frankly *anyone*, for that matter, is when I spilled a cup of coffee in the center console of his truck. The man verily exploded. His normally cherubic face turned bright red and downright demonic. He frothed at the corners of his mouth.

"YOU IGNORANT MOTHERFUCKER I AM GOING TO KILL YOU," he explained.

And he *meant* it. He went so far elsewhere I didn't even recognize his voice. I made myself very small in my seat as we drove, Jimmy fuming, across the desolate Washington Palouse for two hours before he finally said. "Shit, you should know better than that." That was the closest to some kind of half-assed apology he could sneak past his anger.

Other than that, he was the man. He was the one you wanted running your site. Whereas men who ran with Devil Anse and Sarge were constantly begging to be reassigned, everybody wanted to roll with Jim, including me. Jimmy lived on the Missouri side of Kansas City, just a few miles dead east of Arrowhead Stadium, and that is what prompted me to say,

"You know a guy named Jerry Case out of K.C.?"

And he did.

SEPTEMBER 29, 2013

A 25-year-old tower technician (Mitchell Ray Morgan) died Friday morning in Indianapolis when the crane he was operating tipped over and crushed him, according to Indianapolis Fire Department Captain Rita Burris.

The accident happened in a staging area of B-MAC Wireless, in the 8900 block of Rawles Avenue, according to Burris.

Initial reports indicate that the man was moving a five thousand–pound generator but didn't put out the crane's outriggers.

Witnesses said other workers told their coworker to use the outriggers, but he didn't do it, thinking that the generator wasn't heavy enough to tip over the crane.

At some point the operator saw that the crane was starting to lift off the ground, and he tried to leave from the exterior control area and ran, but he was unsuccessful and was killed after being crushed by the crane.[16]

---

16. *Wireless Estimator,* "Tech Killed When Crane Flips Lifting a Gen Set," September 29, 2013. wirelessestimator.com/content/articles/?pagename=Wireless-Construction-News-2.14.

CHAPTER FOUR MISSISSIPPI
THE MISSISSIPPI RULE

Having worked in almost every aspect of the tower industry for over three decades, Jimmy Tanner had crossed a lot of paths with a lot of people. Tower workers are, despite their rough edges, a very elite group of specialists numbering approximately eighty-five hundred to ten thousand employees nationwide.[17] There are more members than in the Writers Guild of America, the only union to which I have ever belonged, and of which I am currently a member in "bad standing." Jimmy did not know Jerry Case well, but he had met him on several occasions while with CES and described him as a sober, hard-working, and very experienced man.

"He seemed to run right and tight," said Jim.

Seemed was the operative word in Jim's assessment because whenever there is a tower death there are nine sides to the story. There is the initial police report, the EMS report, the coroner's report, the OSHA report, the inevitable bevy of lawyer's reports, speculation, erroneous blogs and posts, the story from the crew

---

17. Nobody can agree on the exact number of tower technicians currently in the field. And that is problematic in itself.

on-site, and eventually, *maybe*, the truth. And Jimmy wasn't one to speculate. We joined up with Scotty Hamisch in their satellite office inside the Extended Stay, and I dug up what I could online. There are many Internet sites devoted to the tower industry, but by far the most reliable, comprehensive, and up-to-date on all things tower is wirelessestimator.com. From them I learned:

> . . . *Case was the owner of Structural Inspections Inc. of Blue Springs, MO. Keeling was his employee. Douglas County Sheriff Ken McGovern said the men were five hundred to eight hundred feet high, riding in a man-bucket[18] running along one of the guy wires supporting the thousand-foot four-tenant tower owned by TFM Communications, Inc. of Topeka.*
>
> *Something caused the bucket to crash to the ground. The sheriff's office is working with the federal Occupational Safety and Health Administration to determine what equipment malfunctioned. Paramedics, who were on the scene when deputies arrived, pronounced the victims dead at the tower site. The men were hired by television station KSHB to install an antenna for NBC Action News. A broadcast engineer who knew Case said the climber was a nationally renowned engineer and tower worker, and a phenomenal person.*
>
> *"There were times when he climbed out on the tower with two, three, and four inches of ice on it in the middle of winter to get us back on the air," said Mike Cooney, director of engineering for Entercom Communications Corp. "He would do whatever it took. And there's the attitude that the tower crews are crazy and he wasn't. He was sane and normal, incredibly intelligent, and very passionate about what he did."*

---

18. *MAN-BUCKET or MAN-BASKET:* A conveyance by which tower dogs are hoisted to the workstation by a crane or winch.

The bodies of the deceased workers were taken to the Shawnee Coroner's Office in Topeka, where an autopsy on each was to be performed.

The posts and blogs started erupting all over the net as soon as the story broke. The one that caught my eye and seemed of some relevance was from peaches_cream:

Jerry Case was my husband's best friend. They'd known each other since childhood. Jerry's father (who passed away several years ago) was in this business before Jerry. They NEVER cut corners when it came to safety. Jerry used only the best safety equipment. Jerry and Kevin were using this safety equipment. They were "attached" to the bucket being raised by a cable. It was the cable that failed, causing this accident. And that's what it was, an accident. Nothing Jerry or Kevin or their operator did caused it.

Case closed, I assumed. Being in a man-basket, a kind of shark cage at altitude, was something we at K.M.C.A. rarely did. We climbed, no matter how high. On the few occasions I was in a man-basket, I was scared. I had put my life in the operator's hands. I felt much safer with flesh on steel when I ascended or descended a tower. In a bucket or man-basket, we would lanyard off to the cage, which was the epitome of useless because if the bucket fell it would be like being attached to an anchor thrown into the Verrazano Narrows. "Least they didn't kill themselves," I said.

"You don't know that," said Scotty, and Jimmy agreed. I didn't get it. And they sensed my confusion. They were used to tower dogs not grasping the obvious.

"If they were in the man-basket, who was running the winch?" Scotty said.

"But it says their operator didn't—"

"*Peaches and Cream* said that," said Jim. "Operator could have fucked up and—"

"And who *OWNED* the winch?" Scotty chimed in.

I could see where they were going. We would discover later that there was indeed no error on the part of the winch operator; the cable just snapped. And cables don't just snap. I could see Jim formulating a scenario in his head.

"A friggin' three-quarter-inch rope can haul three thousand pounds up two thousand feet of tower like it's a bale of straw," Jimmy advised. "These cables just don't *break*. You can lift a tank with these suckers."

It wasn't so much a catastrophe to him as it was a problem to be solved. A conundrum to be clinically dissected.

"If you own the winch, you better make damn sure it works properly, and if you rented it, you better do it *twice*," Scotty said.

"How was the block for the cable on the tower mounted?" Jimmy said. "Or was there an *existing* rigging system installed already? *If so*, when was the last time it was inspected and did Jerry and his crew check it out? Was the cable the right diameter for the load? Was the drum operator certified? When was the last time the cable had been inspected?" Jimmy shook his head and stared at his feet. "The bottom line is that if they would have done their due diligence and load-tested the line, they'd be at the bar right now watching the Royals lose."

"But how often does *anybody* really test a load line,[19] Mr. Tanner?" Scotty contested, tweaking him a bit.

Mr. Tanner didn't blink.

"Not often enough," he said.

I knew Jimmy and Scotty would never say it, but I also knew that in their minds Jerry Case killed himself and Kevin Keeling as sure as if they held hands and took a swan dive down into the Eudora dirt. Sunburst Tower Systems owned just such a winch that had been lying out in the yard under a tarp at the Navaho Road shop

---

19. *LOAD LINE*: The means, by rope or cable, by which equipment and material is raised to the workstation.

for several months. Jimmy got on the phone to Brody and instructed him to grab any "slap-dick shop mutt" and have him pour several quarts of 10W50 motor oil over the cable and the cable drum.

It was starting to rain. I wasn't going to get on a crew that day, and as Jimmy continued his welding under a pop-up, Scotty and I took a ride into Ramsey proper and sat at the good wood Irish bar of Grady's At The Station, one of the many saloons along the stops dotting the NJT line from there into Hoboken and Manhattan. Jimmy and Sarge were big, but Scotty was taller than both of them by a head and just as thick. Jimmy called him the immovable object, saying that if Scotty planted his feet and hunkered down, *three* men couldn't budge him from that spot.

"He has an unnaturally low center of gravity," Jimmy said.

Scotty sported a full beard and tussles of unruly black hair. Whereas I learned everything I knew about running cellular antennae and line from Brody and Angelo Kilfoyle back in the early days, I learned how to stack from Jim, Sarge, and Scotty. My first tower erection was a 240-foot self-supporting tower[20] in Spanish Fort, Alabama, with Bo and Gunn and a few other guys whose names have escaped me. Scotty ran the job.

At the edge of Ramsey's postage-stamp downtown, as the commuter trains hissed in and out of the station a few yards away at eighteen-minute intervals, Scotty and I caught up a bit. We drank Yuengling, something we didn't find very often in our travels.

"You gonna' stay with your brother?" he asked.

He knew my brother Rob lived in the house we grew up in, fifty-one miles across the Hudson River in Levittown, but what these Southern boys never understood was that a mile in the Northeast Corridor was like ten miles anywhere else. To get from HERE to

---

20. *SELF-SUPPORTING TOWER:* A pyramid of steel (similar to the Eiffel Tower) consisting of three to four legs joined by cross-bracing. The tallest self-support tower in the United States is the twelve hundred–foot red monster in downtown Kansas City. It is a car-killer, an insurance company's nightmare. Each winter thousands of pounds of ice leave the tower and crash onto the streets and vehicles below. Jimmy Tanner was once asked to put in a bid on taking that tower down. He declined.

THERE would require an excursion down Rte. 17 to Rte. 4 to I-95, across the George Washington Bridge, the Cross Bronx Expressway, the Throgs Neck Bridge, Grand Central Parkway or the Cross Island Parkway, the Long Island Expressway, the Northern State Parkway, the Wantagh State Parkway, and finally Hempstead Turnpike, Gardiners Avenue, and home to Abbey Lane. MapQuest said I could do that in an hour, but MapQuest was operating on crack. I could do it in an hour if I left New Jersey at 2:00 A.M., got to my brother's at 3:00 A.M., slept until 4:00 A.M., and headed back to Ramsey. Toss in the exorbitant tolls both ways and such a thing wasn't close to feasible. I hadn't worked out my long-term lodging yet. I would see which tower dogs wanted to share a room because markets like New Jersey would eat up your per diem in no time at all. My single room cost me ninety-four dollars, and current per diem was sixty-five bucks a day.

But my New York home was still tantalizingly close, and thinking about it reminded me that there was a woman who lived on my block in Levittown that I had remained in sporadic contact with over many years. She now lived in Franklin Lakes, New Jersey, and that was just a few minutes from where Scotty and I sat. I made a mental note to give her a shout while I was here because I knew I would be around *for a minute*, as the Southern boys liked to say.

It was pouring outside now, a full-blown rain, and we knew the crews would be headed back. In the tower industry, when the sky falls, the beer pours, so Scotty and I resigned ourselves to drinking all the Yuengling the bow-tied bartender could slide our way. Inevitably, the talk turned to towers.[21]

---

21. *TOWER:* Amid the nomenclature of the industry, there is no such thing as a *cell phone* tower. They are *communications* towers. And at last count there were 247,081 of them speckled about the United States, with more being built every day. There are three basic types of towers: the monopole, the self-supporting, and the guyed-wire. Though different manufacturers offer various models of these towers and certain installation requirements might demand alterations (alterations that are as limitless as the imaginations and vagaries of engineers), their basic framework hasn't changed since Marconi and the advent of radio. Each style of tower, and indeed each individual tower, presents its own unknowable degree of hazard and difficulty for the men and women who have to climb them. Unknowable, until confronted.

For someone who had installed as much technology as I had—cellular, microwave, GPS, 911, omni whips, grid panels, 2G, 3G, yagis, etc.—it was embarrassing how little I knew about how the shit actually worked. I was foremost a top hand, hanging steel, antenna, and line and all the accouterment. I installed the skeleton and muscles of the systems, the guts of which were housed in millions of dollars' worth of radio cabinets in the site house below. The unvarnished truth: that shit was above my pay grade. I was a tower dog, not a field technician. And the world of cellular changes so quickly that by the time you learn it, it is obsolete. Being out three years, I might as well have been Rip Van fucking Winkle.

"Know what 4G means?" Angelo Kilfoyle liked to say. "THAT means you've done gone up the same goddamn tower *four goddamn* times."

Even though I knew most of the internal workings of the systems were beyond my capacity to understand them, let alone apply them, it didn't hurt to try to know the basics so I didn't sound like a frigging neophyte when we sat around the parking lot at night getting pie-eyed. If you asked Sarge how a microwave dish worked, he would explain it in tradesmen's terms, and you'd stand there gaping like a clubbed mackerel. But Jimmy and Scotty would at least try to ease such information into our heads as painlessly as possible. It was Scotty who explained to me the basic construct of what we all know now as cellular phones. He laid out the two bottles of Yuengling before him. At the base of each bottle he placed a saltshaker and a pepper shaker. He took out his phone and made me take out mine.

"Okay," Scotty said. "Your bottle is a tower in New Jersey and my bottle is a tower in Nashville. So you are here and I am there. You call me . . ." He traced his presentation with his stubby finger, throwing in occasional *beep, beep, beeps*. "Your phone talks to the nearest tower in Jersey; the signal hits a receiver in the antenna mounted at 180 feet. That receiver sends the signal down to this saltshaker in the site house, also a receiver. The saltshaker sends the signal to this transmitter, the pepper shaker, and the transmitter sends it back up the antenna *transmitter*, where it shoots it out to the next tower, and so on and so on and so on until it gets to the tower in Nashville, and that tower shoots it to my phone and I don't answer because I see it's you."

He drank his tower.

"That's *it?*"

"That's it."

"But how does it KNOW which towers to talk to?"

Scotty smiled a cryptic smile and leaned in as if he were about to impart the wisdom of the cellular ages unto me.

"*Elves*," he said.

He lit a cigarette and promptly was told *you can't do that in here*, and we took our beers outside, huddling under the awning in the rain. "Elves and maybe a few satellites," Scotty said. But he could see I wasn't giving up the subject. He exhaled smoke and dropped his head.

"You know what a synapse is, right?" he asked, his tone assuming I did. "Think of the cell phone towers as synapses firing in your brain at the speed of light. Millions of actions and reactions a second. See? From the moment you decide to take a sip of that beer, the process is set in motion inside your skull, and within a split second thousands of decisions regarding the operation of your body have already been decided for you and—" He downed his beer and burped. "Voila!"

"But how does it actually WORK?" I asked.

"That," he said with jovial finality, "is *way* above your pay grade."

Scotty didn't know how it worked. Devil Anse and Jimmy didn't know. I knew for sure that Brody didn't know, and I would wager neither did Sarge. Scotty and I closed down Grady's At The Station, and when we got back to the Extended Stay it was pouring and dark. A few dogs I never met were huddled beneath one of the pop-ups, drinking beer out of the cooler. The vacant lot had turned into a shallow pond, and one worker was standing in water up to his ankles, barefoot, doing some kind of tai chi thing in his pajamas. None of my guys were out and about, and I had no idea who was sleeping where, so I decided to crash in the car. I parked in a far corner of the lot, dropped the backseats, rolled down the windows a few inches, placed my hand on my dad's knife, and tried to go to sleep.

I had always found the syncopated drumming of rain on the roof to be the world's most effective sleeping pill. But not tonight. My synapses were firing like a Gatling gun, and I couldn't turn them off. I saw an aerial view of the whole country with cell phone towers blinking on and off like lights on an old AT&T switchboard. At first there seemed to be no logical progression to the images ricocheting inside my head. I was a righty and my mind was making left turns.

"Left turns are *everything*," Tony Bill once told me as we sat atop his rooftop offices down in Venice, California. When an Academy Award–winning producer and director told you things, you listened. "When I look at a script and know exactly what's going to happen on page ten and page thirty and page seventy-eight, you've lost me. I want to be knocked sideways. I want to say, *shit* I didn't see that coming, but of course I should have seen it coming. I don't want to be manipulated—I want to be *plausibly*, *satisfyingly* manipulated." It was Marc Shmuger who gave me my first job in movies, but before that it was Tony Bill who gently shifted me from theater to film. He was a lot like my dad in that he didn't place much emphasis on his past accomplishments. Every year he was invited to attend the Academy Awards, and every year he'd sit it out.

"Why don't you go?" I asked.

"I'll go when I *earn* it," he said.

My father used his killing knives from the Green Berets to stir paint. Tony Bill used his Oscar for Best Picture as a doorstop. His left turns established him as an industry icon, one of the first true independents. *My* left turns were driving me fucking batty.

It started with me thinking about Laurel, my friend from Levittown. Whenever I thought of her, the first image to pop up was a lamppost. It was the lamppost in front of her house. I had known Laurel since we both were five, and that lamppost seemed to anchor the whole neighborhood. We would congregate there after school with all the other kids who lived on Abbey Lane. As we grew older, we assaulted it with graffiti. The access panel at the base of the lamppost was broken, and inside we could leave secret notes for each other, usually regarding some innocuous gossip at the expense of our friends. Laurel and I never went on a date. We never kissed. That was a good thing for her because she ended up with a good man who made a good living, and they had five passable kids and now lived in a house in Franklin Lakes that you could park a couple of buses in. Still, Laurel, synaptically, was *home.*

Synapses.

Laurel *POP* Levittown *POP* Lamppost *POP* hide-and-seek *POP!*

I sat up in the car and hit my head on the roof.

Hide and seek?

What the fuck.

Sleep was a ship that sailed. I got out and trotted over to the pop-up where one tower dog was still sitting. I startled him, and he palmed his one-hitter and smiled a shy bright smile.

"Don't worry about that," I said, and I slid a beer out of the cooler. Natural Lite, possibly the most disgusting beverage ever to disgrace a brewery, but the dogs bought them by the thirty-pack because it was cheap as sand. By the headlights of the cars still cruising down Rte. 17, I could see he was young, mid-twenties, and slight of build with a dark, uneven crew cut.

"I'm Delaney," I said, and I sat in a canvas folding chair filled with water, which instantly crept straight through my jeans.

"Delaniac?"

"Yeah," I chuckled.

"John," he said, extending his hand. "Or J. Or Junior. Don't matter." He fired up his pipe, took a hit, and offered it to me. I declined with a polite shake of the head. He shrugged and hit it again.

"John J. Junior?" I said. "I dub thee Triple J. Who you rolling with?"

"I'm on Devil Anse's crew."

I shot him a look that said, *You have my deepest sympathy.*

"Yup," he concurred.

"So who are *you* related to?"

"Rick. He's my uncle."

"Ricky Boots?"

"Yeah." I had heard enough to take a guess.

"Georgia?" I said.

"Nope."

Triple J, twenty-five, was from a small town in Northern Alabama not too far south of Chattanooga, Tennessee. He, like most tower dogs I knew, had little education, a "baby mama," and a menial job with menial pay and some sort of contraband addiction and minor trouble with the law before his Uncle Ricky got him on with Sunburst Tower Systems. He had only been out a few days and he was loving it. It was the best thing that ever happened to him. He was no different from the thousands of other young men who somehow stumbled into tower work. And that is what they did, stumbled into it. A friend of a friend, an uncle or brother or sister. There are no real trade schools for tower work, no job-specific union halls or headings in the help-wanted ads. We were hiring guys off of barstools and at bus stops and in the Home Depot parking lot. I was a tower dog because I met a wannabe soap opera actor slash stripper by a dumpster on West Twentieth Street. Brody Kilfoyle, fully aware of the lack of experience in new hires, started his own in-house five-day training classes led by Gray Swain. Brody built his *own tower* so men could learn. And one of Gray's opening classroom lines was *Nobody ever got up in the morning and said, Today I am going to climb towers.*

"So how do you like it?" I said.

"Love it."

"Anse will break your balls, but you will learn a lot from him if you can get past the screaming."

"He bought Ricky new teeth."

That was Devil Anse, all right. Ride you like a redheaded step-child eighty hours a week and then lay out three hundred bucks in singles for the Bada Bing! up the road, five hundred for your child support, or thirty-five hundred for your dental work. Anse's mind was closed in a lot of ways but his wallet was always open.

Triple J excused himself and went to bed, leaving me in the shadows with my lukewarm Natty Lite and my half-dry itchy ass as I tried to decipher what the hell *hide-and-seek* meant. I endeavored to wash my mind with thoughts of Meagan and my little boy, how she walked around the house with her robe half open, how his little fingers grasped my ears and nose, trying to decide what they were, and that was good for a few happy moments. I climbed back into the back of the Explorer and tried again to sleep.

At 4:00 A.M., which had somehow become my witching hour, I *understood*. POP back to Laurel and the lamppost on Abbey Lane. During those Long Island summer nights we would play hide-and-seek and that lamppost was home base. And whoever was *it* had to count to one hundred. One Mississippi, two Mississippi, three—

Left turns, Delaniac.

One Mississippi was one second. And in Gray Swain's classes he would preach the Mississippi Rule. If you are on a tower and you for whatever reason disconnect from your PPE, One Mississippi is it all takes for you to die a traumatic and untimely death. That isn't even enough time to *know* you were going to die—to say *oops*. And among my jaded brethren, there was some small consolation in the idea that when you fell it was over so fast you *never knew what hit you.*

But now I applied the math skittering about my brain with the incidents of the last two days. Case and Keeling . . . seven hundred feet . . . 32.2 feet per second . . . one Mississippi . . . two Mississippi . . . 177 feet per second . . . 5.5 seconds until terminal velocity . . . 2.9 seconds until

POP.

Jerry Case and Kevin Keeling had approximately 8.4 seconds to dwell on their situation as they descended to the earth. They KNEW what was about to hit them.

One Mississippi.
Two Mississippi.
Three Mississippi.
Four Mississippi.
Five Mississippi.
Six Mississippi.
Seven Mississippi.
Eight Mississippi.

NOVEMBER 24, 2013

A 27-year-old tower technician died yesterday as a result of injuries he received on Friday when he fell off of a 340-foot self-support-ing tower in Wichita, Kansas. Douglas Klein, who resided in the New Jersey–New York metro area, had been working with a three-member tiger team troubleshooting a carrier's equipment failure on the SBA Communications–owned structure located at 441 East Twenty-ninth Street North.

At approximately 5:30 P.M. he was coming down the tower after the problem was resolved when he fell. Klein was employed by Pin-Point Towers, whose corporate office is in Omaha, Nebraska. The company did not immediately reply to a request for additional infor-mation regarding the incident.

According to workers familiar with the accident, the crew was working for Ericsson on a Sprint Network Vision project mainte-nance call. Klein reportedly had been tied off to the structure's safety climb cable, according to two sources contacted by Wireless Estima-tor, and somewhere at or above the sixty-foot level, he fell.

It is not known if Klein's equipment failed or if he disconnected to get past an intermediate cable guide, but a coworker said that he was attached to the safety climb during his initial descent. When Klein fell he was severely injured after coming in contact with an ice bridge post. Emergency workers transported him to Saint Francis Hospital.

Klein's parents were able to fly into Wichita to be with him before he passed away. Klein, according to knowledgeable sources, had only been working for PinPoint for approximately three weeks and did not have a certificate of competent climber training from PinPoint. Klein's passing was the thirteenth fatality this year.[22]

---

22. *Wireless Estimator,* "Tech Succumbs from Severe Injuries Sustained After Falling from a Kansas Tower," November 24, 2013. wirelessestimator.com/content/articles/?pagename=Wireless-Construction-News-2.14.

CHAPTER FIVE MISSISSIPPI
WHEELS OVER INDIAN TRAILS

"I *can't* hold their hands," Brody once said, with the inflection that he really *wanted* to hold their hands. "I *can't* be in fifty places at once." Brody wanted to be in the field. He missed it. I could see it in his eyes when he paced like a caged cat behind his stand-up desk, arguing with bankers and lawyers. "I'm not sure I want them to fear the tower, but I sure as fuck want them to respect it." He called them his *boys*, and his dad, Big Jim, called them *Brody's island of misfit toys*. And at night, when the urgencies of business slightly abated, I knew he was worrying himself sick. His creation, which had started with the same shoot-by-the-hip naïveté as his acting and stripping and singer-songwriter career, had exploded into a full-blown money-making reality. Like the Godfather, Brody was always well aware of the personal triumphs and foibles of his employees. But the market in North Jersey was begging for hands. "You can't dip your ladle in an empty labor pool," Sarge Schmidt said. "Sometimes you gotta take what you can get and pray for the best." The new faces were brand spanking new, and many of the experienced hands were guys the company had fired two or three times already. K.M.C.A. was not scraping the bottom of the barrel. They were rooting beneath it.

You hear a lot of corporate speak about *family*. The *family* at Walgreens, the *family* at Walmart, the *family* at Smiling Jack's Ford. Why advertisers think we swallow such maudlin bullshit is beyond me, but apparently it works for retailers and politicians because using *FAMILY* seems to be as common a sales tactic as *wait, there's more*. But for Brody Kilfoyle and his brother Angie, who had risen to general manager, second in command, *family* meant treating us with love and respect and worry and no small amount of ball-busting. They had always treated us as family, but there were so many new faces signing up that Brody was feeling a bit out of touch with them, so much so that he had one of his secretaries create a poster with all his employees' mug shots and names so when he ran into a dog he could shake his hand (and Brody had a grip like a boa constrictor), look him in the eye, and use *that* name. His leadership was not avuncular. It was damn near parental. He would dish out admonishments with the caveat that *you know I love you, Delaniac, but you cost me about three thousand dollars yesterday, and I don't even love my wife that much.* He had a way of making even the roughest dogs yearn for his acceptance and approval. The bulge in his payroll had repercussions that were not financial. I am four years older than Brody, and I could see the gray creeping into his full head of blond hair and the crow's feet tracking out from the corners of his twinkling powder blues.

"When your kids are little, you want them to jump into that pool without hesitation," he said. "But you don't want them to *drown*."

Though the industry as a whole wished to present the image of the tower dog to the world as the all-American, God-fearing family man wearing a neatly pressed jumpsuit and halo-ringed hard hat, Triple J was the prototypical tower dog, and by no means exclusive to Sunburst Tower Systems. He was not the exception but the rule.

The majority of the men wiring America at that time and up to this moment are not choirboys. They are predominantly fuck-ups.

Me included.

The morning after I first met Triple J, we mustered out and Sarge assigned me as a top runner to a crew with Ricky Boots and Super Mario. I had met Ricky before, and the persistent drool from the corner of his mouth led me to believe that whatever he was on he was still on. But it wasn't drugs. Just like Crack Baby, it was *low blood sugar*. Low blood sugar meant you either had low blood sugar or you were a recovering junkie eating peanut butter out of the jar with a spoon. Either way, Ricky was clean and a damn good hand. K.M.C.A. had a stringent drug-enforcement policy, and when you were randomly selected, you either pissed in the cup or were fired. The effects of the pot Ricky's nephew had smoked the night before were long gone, though the THC could remain in his system till Labor Day. When the company sometimes asked for volunteers to take the UA, my hand always shot up. Despite my own litany of defects, doing drugs was never one of them. Taking the UA at the doping center meant a few hours' pay for tinkling in a plastic shot glass.

Super Mario was Machu, a walrus-mustached, forty-five-year-old groundhog who did the bottom work (or *civil* work). From high above the compound, watching him hustling about the site, his girth and red hard hat reminded us of the video game. *Boop-BEEP-boop-boop. Boop-boop-BEEP.*

We were slated for an installation atop a three hundred–foot self-support tower somewhere around Bayonne, but the Garmin GPS (who we referred to as the Dash Bitch) had a meltdown and sent us over the Goethals Bridge into Staten Island. Though she completely destroyed the morning, she did so with perfect diction. That part of New Jersey is the strip of decaying unsightliness that gave the rest of the state its unwarranted reputation for being the outhouse for NYC. And Staten Island was, well, Staten Island.

Our crew leader, Daryn Baylor, a musician and death metal headbanger from Oregon, was driving. Daryn had just enough faith in his navigational abilities, despite my objections, to motor us all the way across Staten Island, and an hour later we were over the Verrazano-Narrows Bridge and into Brooklyn. Amendments to the *one mile equals ten miles in the corridor* rule were that rush hour in the five boroughs lasts fourteen hours and that, if you're lucky, the opportunities for making a U-turn in a big diesel dually pulling a thirty-two-foot box trailer were intermittent at best. Knowing we were royally fucked for at least another two hours, I resigned myself to being a smart-assed tour guide for my gawking brothers in arms.

Is that the Statue of Liberty?
No. It's a copy.
What are all those trees?
That's Battery Park. Duracell has a factory there.
Where was the World Trade Center?
They have the wreckage on barges parked in the Gowanus Canal. They're going to rebuild it from scratch when they find all the pieces.

Bouncing like backup singers along the deeply moguled Brooklyn-Queens Expressway, we passed the Williamsburg Clock Tower. I was born not far from there in Park Slope, but I didn't say anything. It would be too hard to explain. I persuaded Daryn to exit on Atlantic Avenue so we could backtrack on the BQE, but by the time he found a place to turn around, we were crawling bumper-to-bumper through Bedford-Stuyvesant.

By that time Daryn was past frustration and exorcised his road rage by blasting guttural death metal incoherencies over the speakers, his head punctuating the downbeats like a dashboard puppy with a mullet. He couldn't hear me when I told him he was somehow *now* heading toward the Queens-Midtown Tunnel and that, as the Godfather would say, *we might want to rethink this one.* My two-hour estimate on making it back to the job site in Bayonne just doubled. Later Sarge would say, almost in wonder, "*It took you four fucking morons four fucking hours to go forty-one fucking miles?*"

When Daryn handed him the receipts for the $64 in tolls and $130 in gas we had amassed, Sarge just deflated, dropped his head, and walked away. Though vexed, some part of him admired the zeniths of stupidity his charges could attain.

The Queens-Midtown Tunnel is really *two* tunnels in one fat-ass tube. Two lanes into Manhattan and two lanes out. During the morning rush, Port Authority would adjust the flow to three lanes in and one lane out, and flip that for the afternoon exodus. It was after 9:00 A.M., so we sat on the Pulaski Skyway over Long Island City, idling in five lanes of pissed-off stagnation funneling into one lane of tunnel three miles away. "Like five pounds of shit in a one-pound bag," Super Mario concluded. The Pulaski Skyway would be a major crossing in any other city in America, but in NYC it was just one big ugly hump of a road having one claim to fame—they used it as the Yellow Brick Road when they filmed *The Wiz* with Michael Jackson.

The entire Emerald City stretched out before us left to right. I knew we would be there *for a minute.* I had crossed this bridge countless times on my way in from the John Grace pipe shop in Bethpage to the Working Stage Theater Company on West Twentieth where I met Brody almost twenty years earlier. All these odd connections got me wondering if there was indeed an intangible cyclical nature to life, some mystic sojourn we are destined to negotiate without a Garmin, a wild ride both Ferris wheel and bumper car that you couldn't corporeally escape. I usually never delved into deep-thoughtery, but sitting shotgun, being pummeled by Behemoth, my brain had no place else to go—until I glanced over at one particular section of the overpass. There on the water-stained beige concrete were four words stenciled in peeling and chipped black paint, words painted years ago and still discernible beneath the decades of graffiti. It said

WHEELS OVER INDIAN TRAILS

and that got the left turns going for the entire time it took us to get to under the East River, across and downtown to Lower Manhattan, to Canal Street and the Holland Tunnel, under the Hudson, and back within striking distance, Dash Bitch willing, of Bayonne.

I had seen those stenciled words many times, but I always thought they were some sort of low-budget, poorly conceived campaign for NYC Parks and Recreation, recognizing Long Island's and New York City's Native American roots. It was and it wasn't. In 1626, before Peter Minuit, acting on behalf of the Dutch East India Company, purchased the island of Manhattan from the Lenape Indians for sixty Guilders, or about seventy-two bucks in today's dollars, the natives of the five boroughs were doing just fine. Over millennia they had established an intricate and practical network of trails and fords from Montauk Point to the Pulaski Skyway. The Rockaway, the Massapequa, the Shinnecock, the Canarsee, and the Matinecock all carved their routes throughout the glacial moraine, these routes based largely upon existing game trails toward water. Some of these paths, usually no more than two or three feet wide, were exclusive to a specific tribal territory, and some of them were byways shared by all. But by 1741 only about four hundred natives remained on the islands, and the Dutch and later English, knowing a good thing when they tripped over it, used the same paths to interlink and lock their settlements. Those lanes grew into dirt and clamshell and gravel and macadam and paved roads and highways, into carriage lines and railway lines and telegraph lines and telephone lines. They formed not only the infrastructure for water and gas and steam and electrical utilities but *all* future migration and urban and suburban development and trade radiating from the hubs of New York, Boston, and Philadelphia harbors and across the nation for the next 450 years. And as we lurched our way into the soot-stained maw of the Midtown Tunnel, choking on diesel and gasoline fumes, I realized not much had really changed.

I have never been on a cell phone tower that was not part of an integrated system of towers strategically built along existing avenues, interstate highways, railroad lines, pipelines, and especially power lines. It was preordained. WHEELS OVER INDIAN TRAILS had become the skeletal basis of modern cellular communication.

Later, when field producing for NBC, I would share this revelation with the executive producer assigned to the tower dogs story, and he looked at me as if I had panda bears shooting out my ass.

"You can be very oblique," he said.

Up until then I had only been called "oblique" once. That was when I was going for my MFA at McNeese State University in Lake Charles, Louisiana. The man who called me "oblique" was Richard Wilbur, then Poet Laureate of the United States of America. Actually, he did not call me "oblique." He said he wished I were "less oblique." Either way, I was deemed somewhat "oblique" by the man who knew, as so recognized by Congress. But as we finally pulled up to site in Bayonne, I wasn't feeling oblique at all. It was not quite an epiphany. That would come later. But I did think I had a legitimate and tenable theory, all inadvertently triggered by Tony Bill's left turns and by one John Fekner:

> . . . an innovative multidisciplinary artist who created
> hundreds of environmental and conceptual outdoor works
> consisting of stenciled words, symbols, dates, and icons
> spray-painted in New York, Sweden, Canada, England, and
> Germany in the 1970s and 1980s. A seminal figure in the
> Street Art movement, Fekner participated in recent urban
> art exhibitions such as Wooster Collective's 11 Spring Street

Project in 2006 and in "Art in the Streets" at the Museum
of Contemporary Art, Los Angeles in 2011. Art writer Lucy
R. Lippard writes, "Fekner does in public what a lot of art
world artists don't even do in galleries: he dispels ambiguity
by naming his visions, his viewpoint."[23]

Whatever the fuck that means it was Fekner who, in 1979, climbed up onto that concrete trestle with his stencils and his spray paint and scribed WHEELS OVER INDIAN TRAILS. That message greeted commuters until Earth Day 1990, when he felt he had made his point and painted over it. *Then how did I see it?* If Fekner decommissioned that work of art seventeen years ago, how did I see it? It had to be a trick of memory. Some mnemonic sleight of hand—*pick a bridge, any bridge.* I had seen it there so many times before I just expected it to be there. I *wanted* it to be there. Therefore, it was.

We were geared so low as we pulled up and out of the Holland Tunnel into Jersey City that we belched dense black clouds of diesel exhaust into the faces of the drivers behind us. One of them, who for the last twenty minutes had nothing to look at but our Tennessee license plates and Confederate flag stickers, gunned it around us, rolled down his window, and said, *Fix your muffler you hayseed cornpone fucks.*

"How did he know our names?" Ricky said, new grill sparkling.

The hayseed cornpone fucks pulled up to the job site not four, but five hours late and none the wiser. It was an overgrown gravel lot, strewn with old plastic bags and 40 Dogs and used condoms and bloody needles in the center of an abandoned warehousing district. I did not know if we were in Bayonne, but it sure smelled like it. "Plenty of daylight left," said Daryn, and we unlocked the site gate to the compound, dropped the back door to the trailer, and went to work. While Daryn and Super Mario would roll out the spools of coax and begin the assembly of the new antennas, we would be fly-

---

23. Wikipedia on John Fekner: en.wikipedia.org/wiki/John_Fekner.

ing, hopefully before dark, Ricky Boots and I broke out our climbing gear. "Who's gonna' rig?" said Daryn.

"I will," I said, and I immediately regretted saying it.

Ricky scanned the tower with knowing eyes, patted me on the shoulder, and said, "Better you than me."

"What's the rad-center?"[24] I asked. Daryn grabbed the site folder off the dashboard and leafed through the pages. "*Ahhhh—rad centerrrrrr*—looks like 225 feet."

This first climb in a long time would be hell, and I wanted to get it out of the way. Looking up at the three hundred footer, I knew I would have to rig at least forty feet above the rad-center, so I would be going all the way to the top. That's like doing about three hundred pull-ups carrying my own 182 pounds, my thirty-five-pound harness, a four-pound rope block, a ten-pound twenty-four-inch adjustable wrench, and the rope, which would increase in weight every foot I climbed, until for the last one hundred feet of tower it would weigh about one hundred pounds. I also pouched a Motorola hand radio, two warm bottles of Coca-Cola, my phone, and my digital camera. Essentially I would be hauling almost two of me. I would not be the first man to repeat the phrase *the hardest part of this job is getting to work*. Because it was. The rest of the tools and material and equipment would come up attached to the load line or in canvas nose-bags once I rigged the tower and sent the ground end of the rope, weighted by the adjustable, down to Super Mario. He would then either dog it off or spool it onto the cathead[25]—the truck-hitch-mounted capstan winch. But that would be a while because this was a *dirty* tower in that it was loaded with carriers.[26]

---

24. *RAD-CENTER*: Radiation center. The elevation at which the center of an antenna is mounted to the tower. Also called center-line.

25. *CATHEAD*: An electric winch through which the load line is fed to hoist equipment and material up the tower. This is mounted either to the tower itself or to the rear of a truck with the appropriate hitch/receptor. This term is an old seaman's term, referring to the scream the winch makes when pulling up the load.

26. *CARRIER*: A company that provides telecommunications services. Verizon Wireless, Sprint, AT&T, Alltel, Nextel, and T-Mobile are some examples of carriers.

Between eighty and 260 feet there were no less than six antenna arrays on this tower representing Sprint, Verizon, AT&T, Nextel, and *godknowswho* else. And it was not just cellular hookups. There were three-foot- to ten-foot-diameter microwave dishes and high-performance dishes, grid panels and UHF and two-way radio, omni whips and bi-poles and two old fiberglass feed horns that resembled cornucopias (only they were twenty feet around, so spacious you could live in one).[27] There were beacons and sidelights and what I think was an air-raid siren, and all of it, ALL OF IT, was bleached white with crusted pigeon shit. This beast[28] was as dirty as the lot it was built on. A tower, after all, is just vertical real estate, and carriers could rent as many feet of tower as they needed. Many companies did indeed have their own towers built exclusively for their own systems, but this market was the Wild West, and they were hanging shit everywhere. Towers were suspiciously overloaded with antennas, diplexers, and TMAs, as well as the tons of coaxial cable needed to connect these units to the world below. Site-acquisition agents scoured the tri-state like hopeful ants, renting space on rooftops, church steeples, smoke stacks, water towers, billboards, silos, flagpoles, bridges—if there was a clear line of sight between two elevated structures, there was a good chance they were "talking" to each other. And despite the carrier's claims, despite what they told the concerned neighborhood committees who would have to stare at these aesthetic intrusions for the rest of their lives, no amount of stealth technology could mask their indiscreet existence. You can hide cell phone antennas behind faux brick shrouds, you can paint them, you can dress them in polyurethane branches with leaves, you can affix to them the sign of the trinity or an ad for Chico's Bail Bonds, but they didn't fool anybody.

---

27. On a site in East St. Louis in the winter of 2009, we found a decommissioned horn lying out in the woods. Upon closer inspection a homeless man stuck his head out and said, "Y'all won't be needing this back, will you?"

28. *A BEAST*: Any tower that exceeds the usual degree of difficulty. Example: "I have to climb THAT beast today?"

I wasn't fooling anybody either. The climb was tough. There was a ladder but no safety climb,[29] and I would have to crabwalk the entire height, using my two fall-arrest lanyards (or *pelicans*) to clip on to the steel.

Clank, clip, climb.
Clank, clip, climb.
Clank, clip climb about 150 times.

Though when I first started in the business we would free-climb all the time, crabwalking was now as close as we were allowed to that insanity. It also expended three times the energy needed during a *clean* climb. There were other problems. Some jackass crew before us decided to use the climbing ladder to support four lines of 1⅝-inch elliptical cable, and I could barely get my hands around the rungs. At every grasp, finely powered pigeon shit enveloped my head and shoulders. There was no rhythm to it, and when I did finally fall into a groove, my Petzl would clack against the steel above me as I ran smack into a boom gate. When I reached an antenna array, I would have to settle into my harness and figure out just what gymnastic contortions I would have to undertake to get over them. I could usually do a clean three hundred feet in eight to ten minutes. Devil Anse could do it in five. By the time I got to the fourth array, I was struggling, almost hanging upside down when the radio crackled on.

"How 'boutcha, D?" Daryn said.

What he wanted to say was *Why the hell is this taking you so long?* But we never did that to each other. He was no stranger to hard climbs, and though he was impatient for me to get this baby rigged, he would never rush me. I seldom checked my time on a tower, but

---

29. *SAFETY CLIMB:* A stainless steel cable running the entire height of the tower to which climbers attach a cable grab. The grab is a device carabinered onto your chest-mounted D ring that will allow you to move freely upward, but will lock into place during sudden drops. If you are attached to a safety climb correctly, the maximum distance you may fall is about six inches. Every tower in America should have one. Many do not. And men who have fallen those six inches have soiled themselves.

I had to have been climbing for almost an hour. My biceps were on fire, my hands were cramped, and the sweat on my face mixed with the pigeon shit felt like pancake batter.

"A little rough," I said.

"Copy that," he chuckled. "Belt off and stand by."

Something was wrong.

I shifted back into my harness and looked down. I hadn't noticed earlier, but they had only busted out one of three antennas and were all standing around it, heads bowed, as if in prayer. That was our two worlds. When you work the top, you seldom know what is going on downstairs and vice versa. Top hands and groundhogs were breeds apart in both skill set and body type. Whatever it was, I didn't give a shit, and I rocked back in forth in my harness, using the balls of my feet on the steel as a pivot point, taking in the view, which was spectacular. Tankers and tugs churning up the waters of the Port of Elizabeth, New York Harbor, and the Hudson. Ferries boiling in and out of Ellis Island and the Statue of Liberty. The Verrazano and the GWB sagging under the weight of relentless commerce. The forest of red-striped refinery emission stacks on the Jersey shore. And, of course, the incomparable NYC skyline.

What men can build, I thought.

Outside of Atlanta in the '90s, I rarely worked in cities. Atlanta, so enamored with itself, didn't have a skyline. It had buildings. But to the Southerners I worked with, it was Hong Kong, Paris, and Mecca all rolled up in one. Normally we plied our trade in America's hinterlands. Inaccessible mountaintops, decaying bergs and tank towns, vast expanses of prairie and desert—miles from nowhere. I have been so high in such nothingness that I have seen the curvature of the earth, have seen oil rigs fifty miles out in the Gulf of Mexico, and have been blinded by the yellow smoke of Arizona burning while on a tower in Dodge City, Kansas. That was one of the only true perks of being a top hand—*nobody sees what we see*. Even poor Bayonne (which I had heard maligned since I could walk) didn't look so bad from up here. And I'll be damned, *is that a golf course? Bayonne has a golf course?!* Gentrification rules. Outside of our present little patch of crap, Bayonne did not look all that bad. It had a wonderful view of Lower Manhattan, and Bayonne didn't make that stink—the adjacent acres of refineries did. The same row houses and brownstones were going for millions over in Brooklyn and up in Hoboken. They could even have a Starbucks down there crammed with playwrights and novelists. Breslins with backpacks pounding the urban beat. Little Hemingways, sticking pencils in one ear and quoting Nietzsche. Why not? If I could have a movie made, limited release, anything was possible. Anvil-shaped thunderheads were looming out over the Atlantic, and far to the north by the NY-Jersey line, but they were hours from erupting anywhere near us.

"How 'boutcha, D?"

I *whooped*,[30] not keying up the Motorola.

"Near as I can tell we either got the wrong antennas or the wrong tower," Daryn said. I got on the radio and offered an uncertain, "*Okayyyyyyy.*"

"The specs say the rad-center is at 255 feet, but the label on the antenna says 420," Daryn said.

"That might be a tad difficult," I said. "I forgot to bring my fairy dust."

"Copy that. Stand by."

Down below they moved in circles kicking up dust in animated discussion, and I knew the conversation went something like this:

YOU tell Sarge.
No. YOU tell Sarge.
YOU tell Sarge.
No. YOU tell Sarge.

---

30. *WHOOPING:* Radios can be notoriously undependable. Whooping is a way to communicate. One whoop usually means UP or YES. Two whoops usually means DOWN or NO. I have been told my whoops lack authority, and I sometimes use a whistle. Nobody can whoop like Southerners. It is in their blood.

Daryn came back on the radio.

"Tell me it looks like rain," he suggested.

"It looks like rain," I said, knowing the day was done. Ricky and I snapped the line to the gravel, and I began my descent as Super Mario started loading the trailer. Daryn would explain to Sarge later that *we almost got the tower rigged but thank God we didn't because there is something wrong here because the rad-center and specs don't match and we didn't want to take a chance on wasting any more time and . . .*

At about one hundred feet I noticed some movement amid the warehouse shadows about half a block from Daryn and the dogs. Three figures, hoodies down over their eyes, pants at their thighs, walking in a *keep on truckin'* bop and lunge. I got on the horn.

"Watch your six."

When I hit the ground Daryn was already in conversation with these skells, who couldn't be more than fourteen or fifteen. They were startled by the sight of me and huddled and laughed. I couldn't see the pigeon shit melting off my face like weak chocolate milk. There was no doubt in our minds that these kids were casing our site, looking to rob us, copper being a large part of what we do and a large part of what they steal. But we had been down that road before and knew how to avoid any tension or violence. "Everybody's gotta earn somehow," Devil Anse was fond of saying upon discovering a site torn apart by copper thieves. "More work for us."

"Look guys," Daryn was explaining to them. "It'll take about three days for us to finish this site. After I take *all* my pictures and get *all* my shit out of here . . ."

Daryn then shrugged as if to say *whatever happens, happens.*

"Deal?" he said.

The three boys stepped back and conferred. After a moment one of them thrust out his chin, his lower lip covering his top lip. He cocked his head to the side and nodded the tiniest nod.

"Word," he said.

In the truck and pulling out of the site, I had to smile. Daryn was an outlander, but the street was the street and he could be smooth as mercury.

"More work for us?" I said.

"Word," Daryn said, and he smiled, turned up Lamb of God till the windows rattled, and motored deep into the sputtering flames of the forest of two-hundred-feet-tall Elizabethan gas-release valves. I started to doze with my head against the windshield, and all the magic and majesty I had bestowed on the day and the sights and four pompous stenciled words evaporated into the stench. And reality fell upon me like sleet.

I was a fourteen-dollar-an-hour cog in a thankless deadly wheel (risking my life daily so some fucking tweenie in homeroom class could text *I'm bored* to some other fucking tweenie sitting three rows back) rolling steadily, blindly onward.

Destination nowhere.

JANUARY 31, 2014

A tower technician was killed on Friday when he fell off of a thousand-foot guyed tower located south of La Feria, Texas.

Cameron County Sheriff Omar Lucio said Ronaldo Eduard Smith, 62, was providing maintenance on a tower that belonged to a church when he fell.

Deputies located the man about thirty feet from the tower with serious head trauma following the fall. They did not know how long he had been dead.

Communication between the man and his company was reportedly lost throughout the day, and his office became worried and contacted authorities, who found the man dead at the scene.

Lucio did not have the name of the company that employed Smith.

Wireless contractors maintain a safety policy that requires a minimum of two men must be on-site whenever there is elevated work being performed.[31]

---

31. *Wireless Estimator,* "Texas Tech's Death Raises Concerns About Company's Safety Practices," January 31, 2014. wirelessestimator.com/content/articles/?pagename=Wireless-Construction-News-2.14.

CHAPTER SIX MISSISSIPPI
WON'T YOU TRY EXTRA DRY RHEINGOLD BEER?

Before Tammany (sachem of the mighty albeit guileless Lenape) cut his deal with the Dutch, the natives were doing just fine. Their holdings ranged from present-day NYC across the East River into Long Island, all of New Jersey, tens of thousands of acres in western Pennsylvania and southern New York, and snippets of Connecticut and Maryland. Though it only took about 115 years for the Dutch, Swedes, Germans, English, and Welsh to buy out, push out, breed out, and wipe out entire conclaves of indigenous peoples and knowledge, their hubris and greed would later bite them in the ass but good. Because what the Lenape and the thirteen other nations in the vicinity knew that the Dutch, Swedes, Germans, English, and Welsh did *not* know was where to put things. And they would not have put a tri-state megalopolis in a flood plain. I am not talking about Hurricane Sandy and tidal surges. It only took a couple of inches of plain old rain to render dozens of major arteries throughout eastern New Jersey impassible. Four to five inches and whole neighborhoods were up to their stoops and basement windows and bathtub Madonnas in water. If Tammany could see that today, sitting tall on an outcropping atop the Palisades, his blanket over his shoulders and his small fire crackling in the damp, he would shake his wizened head and say the Lenape equivalent of *morons*.

The heavens had descended as we surged up Rte. 17, through the mall-smothered towns of Ho-Ho-Kus, Waldwick, and Saddle River, filthy water up to our wheel hubs. When we arrived at the Extended Stay, most of the company was already back and the seeds of one helluva debriefing were being sown. Heat was shimmering from the stainless steel lids of the gas grills and smoke chugged from the charcoal grill as three more pop-ups were being erected. Diesel trucks were idling, doors open, all stereos tuned to Q104.3. I could tell Vic with a D was in charge of the music because it was classic rock, the one thing he and I ever agreed upon. By midnight the truck radios would be turned down and half a dozen dogs would break out their mistuned axes and Gorilla amps and play and sing till dawn or till the cops came. Dogs were unloading twenty-pound bags of ice from truck beds to coolers along with cases of beer—piss water Natty Lite and Milwaukee's Best.

This was about volume, not quality.

The floorboards of their trucks held the pints and quarts of Stoli, Captain, Rumplemintz, Wild Turkey, and Beefeater in paper bags. The coveted mason jars of moonshine up from Pigeon Forge in the Great Smoky Mountains would eventually be presented in ceremonial deference and ritual. I did not see any food. That was Sarge and Devil Anse's thing, and they would not disappoint. I knew they were at the nearest meat market, enthralling the butcher with their knowledge of marbling and texture, confounding him with their required specs for each cut. "I got eyes like a micrometer," Sarge would warn. "And THAT chop is NOT ⅞–inch thick." Sarge and Devil Anse did not buy their meat off the rack.

Beneath the ponderous rain, the dogs laughed and whooped and kidded and played. Frisbees sliced and wobbled; hacky-sackers gyrated in a spastic circle. The guy who did Tai Chi in his pajamas was now riding a unicycle, and Cady, a new hire from Daryn's neck of the woods, showed off his considerable prowess on the skateboard. The dogs had landed, and the parking lot at the Extended Stay resembled a playground for insane children.

It was payday.

Glory be.

I took a quick survey and realized there was nobody for me to room with. The dogs were staying two, three, and four to a room, some grown men sharing beds. Super Mario offered me a piece of floor. Bo and Gunn said they'd have a cot hauled in, but between Gunn's snoring and Bo's sleepwalking, I knew I would be better off in the Explorer. Super Mario offered up his shower, and I grabbed a six-pack of M.B. from Crack Baby's cooler, went up to the third floor, and took a bath instead, washing the day away, washing away the guano and the sudden, unexplainable sleet of depression. After three beers, the water was a milky grayish white and cold. I had cleansed the heavy, but I was far from clean. I could hear Super Mario counting his Crown Royal bag of quarters on the table, laughing at a rerun of *Who's The Boss?* He wasn't the brightest bulb but nor were any of us. Most of all he was just a sweet, sweet, round old bachelor with a single-wide trailer on a sliver of bottomland in Western Missouri for which he was saving up to have a water meter installed.

While I refilled the tub, I wrapped a towel around me and stepped out into the hall. The air was already thick with the sickeningly sweet smell of marijuana. That smell always made me nauseous, a reasonable explanation for why I was not a full-blown, Twinkie-sucking pothead. Back in the tub I thought about calling Laurel in Franklin Lakes. My Lady of the Lamppost. I had her number in my wallet. She did not know I was in town, and I debated whether or not I wanted her to know. With her I wanted to be what I wanted to be—the writer—not what I had become. Plus I owed her money. I thought calling Meagan would cheer me up, but it didn't. We did not have what any rational person would call a traditional relationship, but we did have a baby, and I genuinely loved and missed them both. She said what she always said, *We are fine and do not worry.* Prompted by Meagan cooing *Daddy? Daddy? Daaaaaa-DEE!* my son chirped, squealed, and spit into the phone.

By the time I joined the rest of them outside, I had consumed four bargain-basement brews, and though I wanted to drink so much more, I could not stomach another sip of that swill. It was not potable. Jim, Scotty, and Gunn sat about a Gott eighty-quart cooler, and that's where I wanted to be. Scotty saw me coming and raised the lid with a flourish. Behold the stash for the discerning tower dog—Corona, Moosehead, Yuengling, Sam Adams, Boulevard, and Dos Equis, which Gunn thought meant "two horses." Jimmy reached behind him and opened a folding chair for me. Gunn had played football for Oklahoma, and his smile was as big as his shoulders.

"Modelo Negro, *mi amigo*?" he said. "Ay-yi-yi."

It was ice cold, dark, and had bite. The first sip hurt my eyes and mercifully expunged the formaldehyde residue of Milwaukee's Best. I sat deep in Jim's Dale Earnhardt chair, an honor in itself, listening to the men exaggerate their days, listening to the unpretentious laughter, listening to Jethro Tull skating away on the thin ice of a new day. The rain had slowed to a drizzle, and Sarge and Devil Anse had loaded the grills with steaks and chops and chicken wings and corn and pots of beans and baked potatoes. Vic with a D, wielding his brush just so, dabbled his own secret homemade sauce onto the wings and chops. You could levitate on the smoky aroma of it all. "Eat meat!" said Devil Anse, holding up a bloodred steak on a BBQ fork. "The West wasn't won on salad!"

The good cold beer went down like cream soda. And I was better. I had been away from most of these men for three years. I had been back at work for a day and half and they treated me like a brother coming down to breakfast. Though we shared the common thread called work, they were all so very different when you took them one by one. Though disputes, both minor and major, happened daily, you did not hold grudges on towers. All dogs were alpha dogs. We had to be. The stupid muthafucker you wanted to kill last night could save your life today and not think twice about it. Everybody was behaving except White Chocolate, another Georgia boy, who was too new, too familiar, too high, too drunk, and too loud. He worshipped all things ghetto and aspired to be a gangsta. I couldn't stand the little pissant. But all in all the kids were all right. Like Brody, I loved them. And like Brody, I worried about them.

"Good to have you back, Delaniac," Jimmy said, and he and Scotty and Gunn raised their beers in a toast. We clinked. And that's when my epiphany, however short-lived, engulfed me like a warm terrycloth robe.

> *I will no longer just SAY I am a tower dog; I will BE a*
> *tower dog, and I will dedicate the remainder of my life*
> *to the ninety-hour weeks and the brutal 110-degree heat*
> *and fifteen-below cold and lousy pay and the palpable risk*
> *and the ten and a half months out of twelve on the road in*
> *fleabag motels from Idaho to Myrtle Beach eating out of*
> *convenience stores and vending machines and sharing motel*
> *parking lots with crack whores, and I won't surf Wireless*
> *Estimator or USA Today looking for bad news and . . .*

Glory be.

I stood up, swooned, half-tripped over Scotty's unusually low center of gravity, and announced, *We few, we happy few, we band of brothers.* Perhaps it was just as well they couldn't hear me over Procol Harem. These were great guys, yes, but asking them to reference Shakespeare was pushing it. But I had made up my mind, and for the next seventeen days I was by far the damnedest tower dog I had ever been, a money-making, time-crunching company man, proving to the veterans that my chops were still finely honed and to the greenhorns that *I'd work your ass into the ground.* If I were still a writer of screenplays, I would condense the next seventeen days into a gripping montage. And call myself Jack. And be portrayed by a young James Caan.

"Fuck you, Devil Anse," said Bo.

"Fuck you, Bo," said Devil Anse.

They were chest to chest, exchanging dissimilar opinions. They stood by the backhoe, having the same damn argument they would have every chance they got to have it. It was over per diem. Devil Anse was a crew leader but also a partner in Sunburst Tower Systems. He was on salary, by choice, making only a thousand dollars a week in actual pay, less than any of us. He was making his end on the far-away other side of us. He never bitched about it, and he worked longer and harder than everybody else. Trying to outwork Devil Anse was like trying to put a ferret in a wet suit. Two weeks earlier, he was home in Georgia having heart surgery, stint and all. His cardiologist demanded at least two weeks rest and then two to four weeks light duty. Devil Anse was back in a week, and Sarge wanted to kill him. Sarge hid Devil Anse's climbing harness and told everyone that *Devil Anse don't climb or you take on the champ*, meaning Sarge and the ass-kicking you would get. Devil Anse wasn't in market fifteen minutes before he took some belt off a greenhorn, saying, *You don't know how to use this yet anyway*, and up the tower he went. You couldn't outwork the wiry bastard, and you couldn't win an argument. Even if you did, you didn't. While up in Boston, we were dangling off a four hundred–foot smokestack about ten minutes from Plymouth Rock. I suggested we go see it. Devil Anse disagreed. Plymouth Rock was in Maryland, he said. I said let's go see it, and he said, *I ain't driving all the way to Maryland to look at no rock.* Though irascible in sticking to his own suspect beliefs, he rarely bothered to argue with us except when it came to per diem. Per diem was on the other side. Per diem nibbled away on his slice of the cheese.

Bo's beef was that sixty-five dollars a day in per diem was fine in Pukefuck, Arkansas, and Stankass, Michigan, but up here it wasn't nearly enough. "I got to room with Gunn 'cause my per diem won't cover a single," Bo said.

"That's Gunn's problem," said Devil Anse.

"You know what BEER COSTS UP HERE? Cigarettes are ten bucks a pack, and a roll of Skoal Bandits is twenty-four dollars!"

"I don't PAY for your beer or your snuff and your pot, Bo. I pay you fair wage and $455 a week TAX FREE! You can't manage your own country ass, don't come cryin' to me."

Cryin' was the ignition switch. Bo puffed up and moved into Devil Anse. Devil Anse puffed right back like a banty rooster. He wouldn't back up from a Brahma bull let alone Bo, and certainly not in front of a half-drunken audience of dogs. "Chest to chest" was imprecise. It was more Devil Anse's chest and Bo's belly button. They rotated against each other, arms akimbo. Sarge, digging into his steak with grim purpose, had one ear on them. He wasn't going to interfere until Bo said, "Come on."

When Bo said "come on" in that South Texas drawl, it was ON, and we all knew it. So did Devil Anse.

"I'm your Huckleberry," Devil Anse said, and they both dropped back two feet and raised their fists. Sarge stood up and splattered his steak off of a POD, and everything stopped. Even Vic with a D had the sense to turn off the music.

"*Enough,*" Sarge advised.

Bo and Devil Anse relaxed their pre-tangle tightness.

"Fuck you," said Bo, smiling.

"Fuck you," said Devil Anse.

Music up. Resume drinking. Settle down now. By midnight the moonshine had been dutifully passed, and Daryn and his headbangers were jamming. Three guitars and a five-gallon-bucket drum set with a baked-bean pot as a high-hat. They really sucked but decided they would form a band anyway. "What are you gonna call your group?" Cady asked, balancing on his skateboard. The kid looked like he was seventeen.

"The Hayseed Cornpone Fucks," Daryn said.

As usual, our debriefings attracted other guests from the Extended Stay, people like us who worked on the road. Mostly salesmen and installation crews of some kind. Guys who built malls, mainly. Stragglers passing by, looking for a hit, a beer, or a job. A few GeoDyne Tech honchos were there as well and joined in. Every now a then a woman would sashay up but after a few moments realize exactly what she got herself into and beat a revelatory retreat.

The vacant lot, newly christened Lake Scum, was rippling in the wind, the effluvia of garbage tossed onto Rte. 17 riding the frothy tides. But the rain had stopped and a splotchy full moon broke over us, prompting the dogs to howl. Jim, Scotty, Gunn, and I had hardly moved from our cooler. We had been joined by two guys who worked at the Olive Garden next door. One, a Mexican busboy who spoke little English, suffered graciously through Gunn's pidgin Española. The other, a white-haired, black-vested bartender in his seventies, loved to talk about beer and all the great beers of his childhood. He was sloshed, and he and I reminisced about the official beers of the NY ball clubs. The Mets had Rheingold, the Yanks had Ballantine, and he thought the official beer of the Brooklyn Trolley Dodgers was Schaefer. He joyously wobbled down his own undulant memory lane, singing at the top of his voice, which surprisingly was a full-rounded baritone.

Schaefer is the one beer to have
If you're having more than one.
Schaefer pleasure never fades
Even when the day is done.
The most reWARDING flavor
In this man's world
For people who are having fun.
Schaefer is the one beer to have
When you're having more than one.

Everybody applauded this. Whistles and claps. The bartender beamed with acceptance. Daryn's band beckoned him to join them, but he just laughed, grabbed a beer, and fell on his ass. Bo and Devil Anse were sitting on a gooseneck, sharing a one-hitter, back-slapping buddies once more.

"Good thing Bo hadn't clicked yet," said Gunn, and Jimmy and Scotty chuckled.

When Bo clicked, nobody was safe. He'd drop his own mother. When Bo drank, which was the only other thing he did besides work, he would reach a point Gunn called *the click*. It was the point of no return. No matter where we were or who we were with, there would be blood. Only Gunn could dissuade him from this. Gunn could actually time it. When the ice storm of the century turned southwest Missouri into a federal disaster area, we were there, at a roadhouse in Carthage. Bo sat at the bar about three feet away from some goat-roper who was doing nothing more than minding his own business. I sat at a pub table watching Jimmy and Gunn play pool. Gunn looked over at Bo, nudged Tim, winked at me, and said *watch this*. Then he began counting down. *Three . . . two . . .* Bo leaned into the goat-roper.

"What the fuck YOU looking at, Daisy Mae?" Bo said, coming off his stool. "Come on."

You take a ring,
And then another ring,
And then another ring,
And then you've got three rings,
Ballantine.

Those archaic jingles plus forty tower dogs again howling at the moon prompted the night clerk at the Extended Stay to finally call the police. They arrived without fanfare, parked by the road, and, hands on mace, slowly walked among us. One was a tall woman with a long red ponytail. The other a short wisp of a guy with an angular face. Both of them wore their paramilitary garb, the half-gloves, two radios, and their Batman utility belts. They were not going to be caught short when the invading hoards lay siege to Ho-Ho-Kus. As they came closer, I could hear the sound of one-hitters plipping into Lake Scum like a strafing run.

Before the cops got to the heart of the matter, Scotty, Jimmy, and Sarge headed them off at the pass. The bosses exchanged hellos with the cops and listened respectfully. Using only hand gestures (the unofficial semaphore of our world both on-site and off), the bosses then instructed us to turn the music down, clean up the plethora of empty beer cans, lock down the PODS, put the guitars away, square away the kitchen. *We're guests here, dammit.* And we had complied, mostly, but no frantic semaphore could stifle the reek of pot or the sound of the vested old bartender crooning away:

There's just one Schlitz,
Nothing else comes near,
When you're out of Schlitz,
You're out of beer.

"Could you *not* do that, sir," said Ponytail. Angle Face put his hand on his Taser. The old man opened his palms and bowed obediently, and there was a collective sigh of relief amid all dogs with warrants. But the old bastard couldn't let it go. This was his stage, and he was not going to disappoint us, his loyal fans.

"Well, whaddawe have here?" the old man said, hitching his pants up to his chest and toddling up to Ramey's Finest. "Uh-oh! We got Secret Squirrel and Dickless Tracy." Then he spat at their shoes. When they were easing him into their cruiser, I heard him singing,

My beer is Rheingold, the dry beer.
Think of Rheingold whenever you buy beer.

It's not bitter, not sweet,
It's the extra dry treat.
Won't you try extra dry Rheingold beer?

Who could not love this? These guys were the reason I stopped reading fiction. For the next seventeen days we had good weather and worked like the dogs we were. Because of the horrible traffic, Sarge had us muster at 4:00 A.M., and we seldom got back before nine or ten at night. We were cranking and banking, so much so that GeoDyne Tech dumped more and more sites on our plate. Twenty towers here. Fifty towers here. One hundred towers here. And new hires were landing at Newark Airport and LaGuardia every few days. Thankfully about once a week, some slap-dick would get the dog at the Newark bus station, one of the seven verifiable gates to hell. My muscles came back. I was right and tight. I would call Meagan once a day, send most of my money home, shower in any available shower, sleep in my Explorer, and be the man Jungle Boy wanted me to be. Then on the evening of July 30, 2007, two things happened:

1. Sarge tried to strangle White Chocolate to death.
2. Crack Baby said to me, *You hear what happened in Louisiana?*

FEBRUARY 2, 2014

Two contractors and a firefighter were killed Saturday when a three hundred–foot cell tower collapsed and subsequently toppled a smaller tower.

The towers, owned by SBA Communications, are located on Murphy's Run Road in the Summit Park area of Clarksburg.

Four contractors—some of whom were tethered to the tower seventy feet above ground—were attempting to reinforce the taller structure when it fell around 11:30 A.M., killing two of the men.

State police on Sunday identified the contractors killed as Kyle Kirkpatrick, 32, from Hulbert, Oklahoma, and Terry Lee Richard, Jr., 27, from Bokoshe, Oklahoma.

A Nutter Fort firefighter, 28-year-old Clarksburg resident Michael Dale Garrett, was killed when the smaller tower fell as emergency crews made a rescue attempt.

Two contractors remained hospitalized late Saturday, reportedly with injuries that were serious but not life-threatening.

"We are deeply saddened by the loss of life and the injuries to the other individuals and our prayers and concern go out to the families involved," SBA Communications said in a statement. "SBA Communications is cooperating fully with any type of investigation being conducted by governmental agencies. We realize the sensitivity surrounding such an unexpected event and have mobilized our team to be on the scene to assist with any information needed."

Summit Park Volunteer Fire Department Chief Brenda Fragmin said rescue operations were underway to free the four trapped contractors when the tower fell on Garrett.

Jeremy Haddix, Nutter Fort fire chief, said the injured firefighter died at United Hospital Center.[32]

32. *MetroNews*, West Virginia, "Troopers identify 3 killed in tower collapse," February 3, 2014. wvmetronews.com/2014/02/03/two-people-killed-in-cell-tower-collapse/.

CHAPTER SEVEN MISSISSIPPI
I AM HE AS YOU ARE HE AS YOU ARE ME AND WE ARE ALL TOGETHER

Gritz Towers of Goliad, Texas, was having a bad year. Not only had they suffered the first industry fatality of 2007, but now, according to Crack Baby, they would claim the ninth. Both in one of my all-time favorite places, Louisiana. The first incident occurred on February 7 in the brackish bayou estuaries jutting down into the gulf south of Houma, about ninety minutes from where Brody and Angelo Kilfoyle were born. In that instance, the tech fell 330 feet from a self-support tower, and I wonder if shortly before he fell he could see, as I had, those oil rigs anchored fifty miles out in flat green water. His name was Brandon Dale Driggers, of Nocona, Texas. He was married to Ashlee, and he had four kids. He died in a place called Golden Meadows. A beautiful name for such a terrible thing. He was twenty-nine.

The second worker, Gerald Waits of Tomball, Texas, was also twenty-nine years old when he fell from a four hundred–foot guyed-wire tower near Centerville in St. Mary Parish. A land of sugar cane and good fishing. His crew was installing a gin pole[33] when weather threatened late in the afternoon and they decided to call it a day. They were going home. OSHA reported:

> *At approximately 3:45 P.M. . . . Employee #1 and two coworkers had completed installing a Jenn arm (tower extension) on top of a tower in preparation for adding another section to the tower. The Jenn pole was in place when one of the coworkers descended the tower because it had started to rain. Employee #1 was instructed to bring the cable line down after putting it in the roller block. He placed the cable in the block and was moving around the tower away from the other coworker, who was stationary at the time. This coworker heard a grunt, looked up at Employee #1, and saw that he was using his elbows and arms to hold on to the wind lacing on the tower. The coworker was halfway around the tower from Employee #1 and watched as he slipped from the lacing and fell 380 feet to his death.*

According to the official OSHA report numbered 310027842, the "degree" of the incident was reported as "fatality." The "nature" of the same report was reported as "fracture." I did not read the coroner's report because in order to do that (in most states), I would have to get the family of the deceased worker to approve such a review, and I am not going to make that call and scratch at that wound just to inform you of what we already know: that the *degree of fatality* of a human being falling from 380 feet is a lot more devastating than in the *nature of a fracture*.

It is here we walk the wobbly plank of *those who wish to remain nameless*, because I have spoken to several eyewitnesses of these

---

33. *GIN (or JIN or JENN) POLE:* A rigid pole or tower section, from twenty to sixty feet in height, with a pulley or block and tackle on the end used for lifting. When used to create a segmented tower or antenna, the gin pole can be detached, raised, and reattached to the just-completed segment for the purpose of lifting the next segment. The process is repeated until the topmost portion of the tower is completed.

tragic affairs, and their reports are far more detailed. One of them, an EMT and first responder, was a young man who had seen a lot of trauma in his line of work. Regarding a multi-worker tower accident in a place I will not name, he said, *To tell you the truth I couldn't tell where one guy started and the other guy ended.*

While sitting in one of the editing bays at NBC Los Angeles finalizing the graphics for the Tower Dog Special, I urged the producers to run a scrolling list of all the men who died doing tower work over the past ten years, their names and ages and hometowns, as part of the closing credits. I also lobbied for a graphic of the United States with little asterisks blinking like that old AT&T switchboard in all the places those men had died. NBC refused, saying it would take up too much space, not realizing that was exactly the whole frigging point.

Fuck them, too.

Bruno Boudrot, a.k.a. Crack Baby, was from New Orleans, and about six hours before Sarge choked White Chocolate blue, Bruno said to me, *You hear what happened in Louisiana?* We were in the truck that morning rolling to a monopole[34] in a shopping center in one of the Amboys. I was exhausted and feeling like shit. I was not hungover, and that was why Devil Anse had rousted me at three thirty earlier that morning. He needed a sober dog to run down to Baby Dolls, the stand-in for the Bada Bing! in *The Sopranos*, and pick up Sarge and White Chocolate. I did, and during the ride back, White Chocolate was so drunk and obnoxious I wanted to throttle him as much as Sarge did and would. But first I wanted coffee.

Bigfoot was driving. We were spread so thin we were operating two- and three-man crews where normally there would be at least

---

34. *MONOPOLE:* The monopole is exactly what the name suggests, a single free-standing pole, rarely over 240 feet tall, built in sections that are either bolted together at flange joints or stacked in sleeves and held together by gravity alone. These are commonly found in high-population areas and are often disguised. A monopine is supposed to look like a pine tree, a monopalm like a palm tree. They seldom do.

four. But Crack Baby, despite his name, was a seasoned dog, and Bigfoot was all you could want in a groundhog. He was in his mid-fifties, from the Spokane area, and looked like a lumberjack—six foot four and broad as a barn door. If, when loading up in the yard, he saw two or three dogs struggling with a 220-pound spool, he'd just walk up, pick it up, and hump it to where it needed to go. He was not showing off. Just impatient. He was an old friend of Jim's, and standing in the parking lot after work, he would survey the scene, sigh with a *what has the world come to* acquiescence, and pronounce, almost daily, "We're all bozos on this bus."

Crack Baby and I had baptized myself in the Atlanta market. He really did have low blood sugar and sometimes would fade on you while hanging in his harness high above Peachtree Industrial. His candy bars saved his butt on several occasions. In his twenties, he was slim and lithe, with wavy brown-blond hair and a partial-Cajun accent that I could listen to all day. We arrived on-site in the corner of a mall complex so loaded with box stores and niche retailers and fast-food joints I instantly longed for the East Kansas space. Sure they had malls in Kansas, but a few minutes in any direction would take you away from them, into the hills or dales or fields or the glorious gravel back roads. Here, box malls and strip malls and mini-malls were inescapable. Here, they *were* the landscape. I could see the historic marker on Rte. 17.

> *Here, on this spot, in the Year of Our Lord 1627, Tammany, Grand Sachem of the Lenape Indian Nation, used the proceeds from the sale of the Isle of Manhattan to erect the very first shopping mall in North America. Among the original tenants were Old Navy, Game Stop, and Bed, Bath, and Beyond.*
>
> (Courtesy of the Bergen County Historical Society and the Paramus Chamber of Commerce)

Crack Baby and I got out, took one look up at the 220-foot monopole, and knew we were in for a fight. We only had to get to 120 feet, but this beast was festooned with seven carriers in sectors so close the tops of the AT&T antennas were touching the bottoms of the Sprint antennas, which were touching the Verizon antennas and . . .

"Who the fuck *walked* this site?" said Crack Baby.

Whichever one of the bosses did the pre-construction site-walk failed to note that the tower was so loaded the only way to get to our sector was if we wore jet packs. We needed a crane or at least the boom truck.

"You call Sarge," said Crack Baby.

"No, YOU call Sarge," said I.

Bigfoot called Sarge.

Sarge cursed himself and said stand by. He had walked the site and several others in the Amboys and, in a rare lapse of memory, had confused our site with a different Amboy. A kinder, gentler Amboy. While Crack Baby and I inspected each other's climbing harnesses, Bigfoot informed us that there was no way we were gonna get a crane on such short notice, so either Gunn or Bo was coming with the boom truck and the man-basket, and we could chill because it would be a minute. We all got coffee at Dunkin' Donuts and tried to find shade, a sparse commodity. It was hot and clammy with no wind at all. We ended up sitting on the expanding asphalt against the truck, listening to Imus, who had just said *nappy-headed hoes* that week and the subsequent debate was getting lively. Half the dogs thought he went too far, and half of them thought not far enough. I was hoping Bo would bring the boomer. I had taken the main[35] before with Gunn at the stick, and he jerked you around like a ten-pound channel cat on an ultralight rod. Most of the guys loved the basket because it spared them the exertion of the climb, but I was not comfortable in it. This is odd because I had ridden the head-

---

35. *TAKING THE MAIN or RIDING THE MAIN:* Securing one's self to the load line and being raised by a cathead or winch to your workstation. This is now illegal except for special instances requiring special equipment, safety plans, and clearance.

ache ball on cranes and boomers and even the cathead many times in the past. Just me, the rope or cable, and the operator, but *only* if that operator was the Godfather, Angelo Kilfoyle, or Bo. Sometimes with Jimmy Tanner, four or five of us would lanyard off to the ball, and he'd fly us up three hundred feet, about as high as a crane can go. Whether it was blissful ignorance or bulletproof youth that made me unafraid I could not say. All I knew was that now I got the willies. Not that anyone would ever know. Maybe it was my kid. Maybe because I suddenly had someone to care about besides my own self-involved ass my pre-climb synapses checked my usual modus operandi. Had I become superstitious? Around my neck I wore a dog tag embossed with my son's name and date of birth. Before each climb I would duck away from the other dogs and give the tag a little peck. We were told over and over again, and I often reiterated the same to newbies, *Trust your training and trust your equipment and you'll be just fine.* Sipping my coffee with Crack Baby and Bigfoot, I pondered if Jerry Case ever said that to Kevin Keeling.

It was two hours before the boom truck arrived, and I was glad to see Bo, Stetson dipped jauntily over his eyes, rumble into the lot. If I had to get in the basket, I'd kiss the baby and get in the goddamn basket, but at least Bo was a smooth operator (the pun both unintended yet unavoidable). He hopped down from the truck, and we set the outriggers on the railroad-tie dunnage, extended the stick, and hooked up the basket. He ran us up. Bigfoot ran the tag on the basket to ease us this way and that.

One hundred and twenty feet above Dunkin' Donuts, Crack Baby and I traversed[36] from the basket to the tower. Bo would bring up our new antennas in the basket when we were ready. We had hours of prep and so did they below before the cut window,[37] two days from now. To demystify and simplify:

---

36. *TRAVERSING:* The act of moving across the open steel from one position to another.
37. *CUT WINDOW:* The crucial time period when cellular operations are shut down so upgrades and maintenance can be performed. During this time calls are dropped and the carrier is losing thousands of dollars. Making the cut window is high-stress and high-pressure.

*What Crack Baby and I would do is rig the tower and
ready the three existing antennas on Alpha, Beta, and
Gamma sector for removal and replacement (decom),
breaking all the weatherproofing and supports and letting
it hang loosely secured to the tower, playing at some
coverage-loss but still playing.[38] Bigfoot and Bo would
assemble the new antennas on the ground, mounting them
to ten-foot-by-three-inch mass pipes along with TMAs and
diplexers and jumpers (an assembly that, depending on the
carrier, could weigh in excess of four hundred pounds) and
ready them for flying. They would assemble the ice bridge,
if required, and prep the portal entries for the coax runs
from the base of the tower and into the site house. Using
the boomer or a load line and block, we would raise the
three new runs of coax to each sector above and terminate
them with the proper connectors at the top and bottom.
If there was a mounting issue, if the existing steel didn't
match up with the new antennas, we would adjust and
fabricate with stock steel we kept with us at all times. We
would then fly the new antennas and secure them to the
tower. During the cut window, the system would be taken
off-line, and calls would be dropped all over the Amboys
while we attached the new lines to the new antennas,
supported and secured the coax from antenna to site
house, turned the system back up, and, God willing and
the levee don't break, hope it worked.*

That is an oversimplification, of course. I did not mention the
scores of procedures this entails: shooting azimuth and antenna
alignments, setting down-tilts, buss-bar mounting, grounding, post-
hole digging, trapeze hanging, cad-welding, load tests, short tests,
system sweeps, weatherproofing, the three hundred site photos and

---

38. *PLAYING:* A generic term for any system that is up and running.

documentation of model and serial numbers, the K.M.C.A. paper-work, GeoDyne Tech paperwork, OSHA paperwork . . .

Every tower was different, and every job trailer was laden down to the springs with the tools, hardware, and material needed to improvise on the fly and get it done. And there was no one way to get it done. We were not building malls. Though the fundamentals remained somewhat static, there was no one way to do any of this, and every boss and every crew leader did indeed do it differently, which made it frustrating for the dogs because what perfectly suited Daryn might send Devil Anse bouncing off the site house walls. We didn't just have to keep up with the ever-changing procedures and technol-ogy, but with the whims and proclivities of Daryn and Devil Anse and Bo and Jimmy and Sarge and Scotty, and GeoDyne CMs, which could change and often did change on a daily, site-to-site basis. In the best of all worlds, like when I first started out with just Brody, Angie, the kin-dergarten classmate, and the painter/musician/dancer, the company *was us*. We were a crew. Once we learned a system, we could antici-pate each other and *what's next* and go for hours working away with-out even talking. Those days were gone. There was too much work and too few dogs, making the majority of what we did on-the-job-training. We were fuck-ups in more ways than not, but we were not stupid. The job required a working familiarity with every hand- and gas- and electric-powered tool there was, as well as knowledge of rope strength and cable strength, of wind load and rigging and knots, of frequency testing and computer skills. You had to be part ironworker, part electrician, part mason, part paramedic, part operating engineer, part plumber, part mechanic, part carpenter, part systems analyst, and part long-haul trucker. You also had to stay alive. In New Jersey, they had a union for every *part* we had to be. Crack Baby and I did this for fourteen dollars an hour with time and a half over forty. And we were in the top third of the scale. Most of the guys were making from nine to twelve per hour. When at the bar, if someone asked us what we did and we said *cell phone towers*, you could bet your paycheck on two reactions. First they would say, *You're crazy*, and then they would estimate you had to make at least thirty-seven dollars an hour. I don't know where that number came from, but it was always thirty-seven dollars an hour. We would respond with *I wish* and leave it at that. We weren't stupid, but we were no Einsteins either.

Einstein wore the same brown suit every day so he wouldn't have to think about what he was going to wear and free his mind for other pursuits. So did Bo. The same blocked and sweat-stained cowboy hat, the same faded blue T-shirt, the same 501s, and the same armadillo cockroach killers. "Where's Gunn?" I shouted down to him.

"Vic and him went shopping!" Bo said.

We had not had a company-wide debriefing since the Rheingold guy, and not only were we due, but Sarge had invited many a big shot from GeoDyne Tech, warning us of the same that morning during muster, and also advising we be on our best behavior or take on the champ.

"So Gunn gets the milk run?" I said, resentful of the fact that Gunn and Vic with a D would spend the day buying groceries and liquor and propane and sporks while I had to actually work.

"Luck of the draw," Bo said.

I didn't really mind Gunn catching a little light duty, but Vic with a D had become the default company errand boy. He was supposed to be a functioning groundhog. He had skills. Skills we needed on-site. His grounding and support work was incomparable. But not since my return to the field had I seen him do anything but go get this and go get that. During these errands, he would stop at four or five places and do his personal housekeeping, eating sit-down meals and chewing up the day while the rest of us were busting hump and eating cold burgers and drinking warm soda. Okay. So be it. Someone had to do it. I just wished it were me. Off-site, Vic was a pleasure to be around, cooking for everyone, making his Vicaritas by the blender full, playing dominoes, and blasting classic rock. But in my mind we had enough new slap-dicks to run errands. We needed experienced men doing what they were paid to do. I let it go until one job when Daryn and I were hanging off the steel at 210 feet for so long we lost circulation in our legs while Vic with a D sat in the trailer all day reading his newspaper and coloring his hard hat with magic markers. There was plenty to do on the ground, but he just didn't do it, and Daryn and I had to cover his ass when we came down. We were there until midnight, at which time I woke up Jimmy Tanner and told him to pull me or Vic off that crew or I was going home. I wasn't going around anybody's back. I had gone through channels. I had barked at Vic to get off his ass and do something, which Daryn should have done, and I had pleaded with Daryn to do what he should have done. Nothing was done. Next stop, the God-father. Jimmy listened to me rant for a moment. He sighed beneath the weight of another complaint, sat down, and cleaned his glasses.

"Vic came to us," he said. "He's pushing sixty, and he's all tore down." Jimmy was almost apologetic. "He can't do the work he used to do, but we are all he's got. We're trying to keep him busy."

"Fuck him," I said.

"I know you guys are pulling his load."

"No shit."

Jimmy shook his head and smiled. He didn't need this. "Back in the day you'd come off the tower and the bottom work was done," he said. "The trailer was packed right and tight, like a supermarket aisle, truck running with the heat or AC on—and there stood Vic. He'd take your belt in one hand and feed you an ice cold beer with the other."

"When you have a three-man crew," I said, "and one of them isn't doing his job, the other two are doing a job and a half. Vic is the guy you used to fire."

"You want me to fire Vic?"

The way he said that prodded my left turns, and I shot to a reed-hidden back road out by Rockaway Beach. I was in the car with Clemenza. I fell into my best Brooklynese.

"Leave the gun. Take the cannolis," I said.

The allusion didn't take. *Fuck yes I wanted him fired.* But I didn't say it. Vic was a legacy. And he wasn't the only one. Brody and Jimmy carried guys long past their prime or usefulness. That was a two-edged sword because though I admired their loyalty, the price of that loyalty was paid by me and Daryn and every other dog who had to carry the water for these guys.

"No," I said, defeated and feeling a bit guilty, grabbing a beer from Jim's mini-fridge. "He can make Vicaritas all fucking day long, but please keep him off my site." Jimmy chuckled and opened a beer.

"Come the day you get all tore down, wouldn't you want me to do the same for you?"

I would. Tower dogs don't have pensions.

Jimmy usually never drank that late, but one wouldn't hurt. "I'll see what I can do," he said. Then one beer turned into three and four. Jimmy had the best stories, and with the right prompting he'd offer some up. That time in Atlanta they were welding and turned a monopole into a two hundred–foot pillar of flame, the coax melting, dripping like lava and shooting like a Roman candle, flaming balls of plastic falling to the interstate below and into the backseat of a moving convertible BMW and setting THAT on fire. The time in Pennsylvania where the residents were so opposed to the erection of a new tower they arrived on-site with deer rifles and shot the hell out of Jim's trucks and trailers, shattering his windshield as he sped away. That time he fell over fifty feet, bounced off the ice bridge, and landed in the snow. The site tech who called 911 to report the death of a tower climber about pissed his pants when Jimmy came around the corner and went back up the tower, not quite certain whether he had broken his back or not. The Godfather had been there. The Godfather had seen it all. That is why it was a bit disconcerting when he shook his head, looked me right in the eye, and said

It's fucked-up up here, ain't it.

It was not a question. It was a declaration.

Bo and Bigfoot busted out the new antennas while Crack Baby and I monkey-climbed all over our sector doing our prep work. Crack Baby amused himself by giving me a play-by-play of some guido shopping below us. This stunad was the real deal, 100 percent genuine North Jersey guido. He wore a shiny blue two-piece Adidas sweat suit with white double stripes down the legs and arms. Blinding white sneakers. A wreath of gold chains hung around his tree-stump neck, tangled in his chest hair. His haircut, short black and spiked, cost more than I made in half a day, and he was fat, lumbering like a sated tick as he went from one store to another. Crack Baby was cracking me up.

> *Just left Dunkin' Donuts . . . slurping on a jelly donut . . . got white powder on his jacket . . . wipes it off . . . headin' into Hardee's . . . got him a big bag of something . . . checking himself out in the window of Batteries Plus . . . decides he looks fly . . . uh-oh, going for the Carvel . . .*

Then Crack Baby shouted in a thick Cajun patois, "Yo, Boudreaux! A little self-control, eh, Cher?!" The guido stopped in his tracks and looked about.

He didn't see us. They never do.

By three or four I was getting sluggish, having been up since 3:30 A.M. Instead of lanyarding off to two points of contact, company policy, I was down to one, frustrated with getting snagged constantly on the tangle of steel and cable surrounding us. Crack Baby was on Alpha[39] and I on Beta. I was cursing the steel, cursing my gear. "You doing okay?" he asked. I told him about having to play taxicab. Crack Baby did not have a very high opinion of White Chocolate either.

"He's just all wrong," he said. "Can't even say *dog* right."

I hadn't thought of that, but he was right. *Dog*, to each other, was

---

39. Most tower/structure-mounted cellular systems provide a 360-degree area of coverage, divided equally between sectors Alpha, Beta, and Gamma.

a salutation, a heartfelt *we are in this shit together* term of endearment. When White Chocolate said it, it came out ghetto and gravelly, a condescension almost, diminishing us. He was no dog yet and knew nothing about being a dog. And worse, he (sporting his ensemble of boxers, wifebeater, do-rag, and thigh-jeans) was always immaculate, like the guido. And anyone, besides the Godfather, who could work all day and afterward look like he was ready for a job interview was not to be trusted. His job description was standing around. Standing around the lot when other men loaded, standing around the site while other men worked, always bullshitting about something. He talked a great job, but he seldom did it. I would guess he was related to Vic with a D, but he wasn't. He was Sarge and Devil Anse's nephew.

By seven, I was shot. I had to get the weight off my legs and arms, and I crawled up to a two-foot-by-two-inch section of clean flat steel. I sat down and lit a cigarette. The blood rushed back into my legs, making them tingle with pain. The grit on the back of my neck felt like coffee grinds. Below, Bo and Bigfoot had finished the antennas and were hauling them into the compound and tarping them up for the night. I thought about how right about now Meagan would be sitting on the porch cuddling our son, a cool northern breeze stirring the wind chimes, the pampas grass gently rustling. The cannas would be in full bloom now, and the bird feeders would be loaded with oiled sunflower seeds and alive with grackles, and cowbirds would be all a-twirl. One of the coon dogs would bay at some imaginary intruder, and one of the cats would brush by Meagan's legs as our son kneaded her breasts, ready to eat. The ancient tin roof of the barn would be creaking and—

Bo yelled, "RADIO!"

I had the Motorola in one of my hip pouches but hadn't turned it on. I did and radioed back.

"How 'boutcha, Bo?"

"Sarge says don't forget the clams."

"Copy that."

I had forgotten. Sarge could suck down clams like an otter. Where the fuck was I gonna get clams? We were by the Jersey Shore but not *that* Jersey Shore. There were no quaint little seaside fish markets and taffy shops in the Amboys that I knew of. As I dwelled on the shell-fish dilemma, I didn't see or hear Crack Baby positioning above me, but I felt a slight tug on the D ring[40] between my shoulder blades.

"*What*?" I said, annoyed.

"Nothing," he said.

Bo gave us a whoop and twirled his finger in the air—sign for *wrap it up*. Bigfoot was already heading back from the 7-Eleven with a twelve-pack. As I rose from my position, I felt oddly unencumbered. I didn't feel the usual resistance to movement from my harness. It was then I realized that I had not tied off anywhere after I sat down. Not one point of contact except my ass on steel. My two pelicans and my positioning lanyard were swinging in the moist, still air. Without knowing it, I had free-climbed twenty feet to my perch and sat there unconnected. The only thing between the pavement and me was Crack Baby, who had hooked one of his lanyards to my back D ring.

"Who are you, my mother?" I said.

He smiled an enigmatic smile. "I am the walrus," he said, "*mon ami.*" We got in the man-basket and I attached myself to it, knowing how futile that was, and Bo, having a bit of fun, swung us out a good fifty feet away from the monopole, letting us dangle there for a few seconds, swinging gently to and fro. Crack Baby sheared the wrapper off a Three Musketeers and crammed it in his mouth. "You hear about what happened in Louisiana?" he said. I had not, and he told me.

"Was he in a man-basket?" I said.

"Gin pole," he said. "Stupid fuck."

I did not know Gerald Waits's name yet. And I did not know he had a MySpace page on which his handle was towerdog_29. And I

---

40. *D RING*: A anchoring point, usually steel, which is sewn into your climbing harness to accept the connection of your lanyard hooks and other carabinered equipment. The standard harness comes with six D rings: two hip, two thigh, one chest, and one back.

did not know there was only one post on his Myspace page from a woman named Theresa Kane. And I did not know there was a picture of her with a child on her lap, laughing, their hands up and obscuring half of her face. And I did not know that post said:

> *I will always miss you, Bubba . . . always! Almost six yrs and*
> *not a single day will ever pass that I don't think of you! xx*
> *Cheers to you *raises tequila* x*
> *Jul 24, 2013, at 08:08 PM, Power's Cross, Leinster, Ireland*

I did not know there would be no towerdog_30.

I did not know as Bo eased us back to blessed earth that if Gerald Waits had Crack Baby watching his back on that four hundred–foot tower in Centerville, high above the whispering waves of summer cane, Theresa Kane, somewhere in Ireland, might not be so sad when the dog days of July rolled around.

But I do now.

MARCH 19, 2014

Pasadena, Maryland. Fire officials said the man fell from a water tower behind the Eastern District police station and a fire station just before 2:30 P.M. Wednesday in the 3700 block of Mountain Road.

The victim was a worker employed by a private contracting company and was working on communications equipment that is atop the tower, which is approximately 180 feet tall.

Paramedics tried to save the man's life but couldn't, and the man was pronounced dead at the scene. On Friday, Anne Arundel County Fire Chief Keith Swindle identified the man as Chad Louis Weller, of Stevensville.[41]

---

41. WBAL-TV, Baltimore, Maryland, "Man Falls from Pasadena Water Tower, Dies," March 19, 2014. www.wbaltv.com/article/man-falls-from-pasadena-water-tower-dies/7085968.

CHAPTER EIGHT MISSISSIPPI
THE WHEELS ON THE BUS COME OFF, OFF, OFF

Brandon Dale Driggers. Jerry Case. Kevin Keeling. Gerald Waits. They comprised numbers 1, 7, 8, and 9 on 2007's hit parade thus far. Drumroll, please.

Number 2: Anthony Shands, 72. Port of Shreveport, LA. March 16, 2007

... the man had climbed the tower to install radio equipment and to connect cable on the 150-foot tower. He had been in radio contact with other workers on the project. The coworkers, who were inside a building, lost contact, heard a noise, and went outside and found the man had fallen.[42]

Number 3: Nicholas Fischer, 19. Mandarin, FL. June 7, 2007

Sgt. Rick Hike of the Jacksonville Sheriff's Office said that Nicholas A. Fischer was pronounced dead at the scene when he fell one hundred feet and landed upon an equipment building.[43]

---

42. *Wireless Estimator,* "Second Texas Tower Technician Killed Is Second Fatality in Louisiana," March 16, 2007. wirelessestimator.com/content/articles/?pagename=Tower%20 Technician%20Deaths%202007.

43. *Wireless Estimator,* "Indiana man dies after falling from Florida tower," wirelessestimator .com/content/articles/?pagename=Tower%20Technician%20Deaths%202007.

Number 4: Charles Moore, 30. Bluffton, SC.
June 28, 2007

*According to Kirk O'Leary, spokesman for the Bluffton
Township Fire Department, "The victim had cut his finger
and was bandaging the cut with electrical tape, and all of
a sudden, he just fell. We don't know if his line got clipped
or what." O'Leary said, according to a contractor on the
ground, Moore tried to grab a guy wire on the way down.
The technician fell past two fellow workers working below
him. He was wearing fall-protection equipment, but it is not
known if he was attached correctly to the tower or if there
was an equipment failure.[44]*

Number 5: Arthur Lee Snelling, 40. Summerville,
SC. July 1, 2007

*Communications workers were shaken by a second fatality
within a week when a Georgia man fell 177 feet to his death
Saturday afternoon . . . was working on a Summerville
cell phone tower near Yancey Street at approximately 3:50
P.M. when he fell to the ground and died upon impact, said
Dorchester County Coroner Chris Nisbet. The coroner said
Snelling was wearing fall-protection equipment but was not
attached to the tower.[45]*

Number 6: Name Withheld. Age Withheld.
Pinola, MS. July 5, 2007

*At approximately 1:00 P.M. on July 5, 2007, Employee #1
was on a crew of five, including the supervisor, that was
installing three boom-gates on a three-hundred-foot-tall*

---

44. *Wireless Estimator,* "Georgia man succumbs following fall off of Bluffton,
South Carolina 480-foot tower," June 28, 2007. wirelessestimator.com/content/
articles/?pagename=Tower%20Technician%20Deaths%202007.
45. *Wireless Estimator,* "Second South Carolina death within a week stuns communications
climbers," July 1, 2007. wirelessestimator.com/content/articles/?pagename=Tower%20
Technician%20Deaths%202007.

telecommunication tower under construction. Using a
cathead to hoist the booms, they successfully installed the
first boom. Employee #1 rigged the hoist line by connecting
a snatch block to the top flange of the southwest tower leg,
connecting a carabiner through a bolt hole on the flange,
and finally connecting the load hook of the snatch block
to the carabiner. The crew then moved on to the second
boom. It had been lifted to the highest point allowed by
the rigging configuration, and Employee #1 radioed to
the hoist operator to stop. To get the boom to the required
installation location, Employee #1 had to use a come-along
to lift it another five to six feet. Before he could connect the
boom to the come-along, the carabiner broke, and the boom
struck Employee #1 on the head, neck, and shoulder, nearly
severing his right arm. He did not lose consciousness, and
two coworkers on the tower tied Employee #1's arm in place
and helped him climb down nearly three hundred feet. He
was transported by ambulance to a local hospital and then
transferred to University Medical Center in Jackson. He died
two days later of head injuries sustained in the accident.[46]

Two more deaths in September and December, in Pennsylvania and in New Jersey, would bring the total to eleven. These men worked for nine different companies of varying sizes and specialties. These men were in their teens, twenties, thirties, forties, fifties, and seventies. Some were married. Some were not. Some had children. Some did not. Some had decades of experience. Some had weeks. And only one of them got to say goodbye (at least, I would hope he did). Though linked by their vocation and the grisly manner of their passings, the only two things they had in common (with the exception of the man in Pinola, Mississippi) was the confidence to

---

46. OSHA Report Summary Nr: 201363017. www.osha.gov/pls/imis/establishment. inspection_detail?id=308776848'.

flout the first and most important rule of tower climbing—*belt the fuck off*—and the conviction they *would not die* that day in doing so. In coroners' offices across America there are thousands of photos depicting brutally broken bodies lying lifeless and bloodied, wearing perfectly serviceable climbing harnesses. "I sometimes wonder why we even issue the damn things," Gray Swain would say.

We know the climber in East Deer Township, Pennsylvania, Daniel Plants, broke two rules. He obviously didn't belt off, and he was also working alone, a massive no-no. Nobody knows when he fell or from how high on the hundred-foot self-support tower. His body was found by a little girl walking her dog, something I'm sure she will carry with her for the rest of her life. But at least she has the rest of her life. That's more than Arthur Crane of Stanhope, New Jersey, the last casualty of the year, had. He died after falling 113 feet in Bridgewater, 42.2 miles from where I pulled into the parking lot at Baby Dolls to pick up Sarge and White Chocolate at three fifty in the morning.

If Baby Dolls had not been used as a location setting for *The Sopranos*, it would be just another strip club on some stretch of dimly lit service road in Anywhere USA. And a bad one at that. Technically it was a bikini bar because, by Jersey law, the dancers have to wear bikinis that they are not allowed to take off. Hell, even in Kansas *live nude girls* meant *LIVE NUDE GIRLS* in all their misdirected glory, by cracky. I was never one for strip clubs. I found them to be innately depressing. And they never really smelled all that good. Like old sex and stale beer, cheap cologne and baby power. Like forlorn hope. For my bachelor party my brother and friends "took me" to a strip club in Suffolk County, Long Island, and while they were changing hundreds into singles, I went two doors down to the corner bar and played video poker until 2:30 A.M. By that time they had all gone home, and I had to catch a cab back to Levittown. They had no idea I was not with them. Since then, the only time I have spent in strip clubs was when I was dragging tower dogs out of them. There is nothing sadder than some tower dog who has just spent six hundred dollars in lap dances leaning into you with whiskey breath and saying *I think she likes me.* When in Ybor City, Florida, I was once awakened in the middle of the night by frantic calls from some strip joint where one of our guys blew his paycheck and then ran up a bill for eleven hundred dollars and was essentially being held hostage. I dumped that one flat in Sarge's lap, the situation being miles above my pay grade. Sarge got on the phone, and some bouncer from the club said, "We have his wallet and his credit cards and his car keys, and you ain't getting him back until we get our money."

"Keep him," Sarge said, and he went back to bed.

Strip club. Bikini bar. The distinction was not lost among the tower dogs, who, upon discovering they were paying tits-and-ass prices for the same erotic twinge they could glean from the Sears catalog, soon tired of the place. But not Sarge. For Sarge, the Bada Bing! became his own satellite office where he entertained executives from GeoDyne Tech almost nightly. These men controlled our destinies, and Sarge was painfully aware of that. We had contracted to do x amount of sites, but that didn't mean shit unless those sites were *released*[47] to us. We were not the only company in market. There were other crews wanting that work. And K.M.C.A. had been burned many times before. In Boston, in Mississippi, in Atlanta, and especially in the Tampa area. There we had mobed almost twenty men to market and proceeded to sit on our duffs for weeks. You don't want twenty tower dogs sitting on their duffs for a few days, let alone weeks. We were ready to take on three or four sites every three or four days, and the contractor was squeaking out maybe one site a week. Devil Anse practically went broke opening that wallet and advancing everybody. "What do you need to get you by?" he'd say. It was there that boredom got the best of us, and we got into bar fights,

---

47. Though a general contractor may have promised you fifty sites in a market, urgency often precludes the filing and finalization of the paperwork permits, purchase orders, and NTPs (notices to proceed) needed to commence actual work. And more than once K.M.C.A. had mobilized crews to market only be to told by the general contractor, "Yeah, I had fifty, but only six are ready to go."

got thrown out of hotels, and got ransomed by bouncers. I got so bored I bought a parrot. So Sarge, at least thrice bitten, did the math in that calculator head of his and learned the Jersey way damn fast. The squeaky wheel does not *get* the grease. The squeaky wheel *supplies* the grease. And Sarge smeared that grease all over the smiling boys from GeoDyne Tech. Expensive dinners. Outrageous bar tabs. Fishing charters for striped bass off of Sandy Hook. Tickets to ball games. These guys never said no, and they never picked up a check. So when you heard Sarge was at the Bing, you knew he was prying out sites, but having a good time doing it. So good, in fact, Angelo Kilfoyle would later tell me Sarge tried to invoice about thirty-five thousand dollars in "miscellaneous expenses" at the Bing.

When I pulled up, Sarge and White Chocolate were about thirty yards apart. Sarge was on the phone to godknowswhoatthathour and White Chocolate was having an animated discussion with some imaginary opponent whose ass he was threatening to kick. They both got in the back. Sarge said "Thanks, Doug," which surprised me because he so seldom used my first name. I was not surprised to see White Chocolate there because he and a few other dogs could smell free drinks and food like turkey buzzards circling roadside carrion. They knew that at a restaurant or bar, if you sat anywhere close to Sarge or Devil Anse, and often Jim, they would cover your tab as well. We all took advantage of such graciousness once in a while, but not every day. It was prideless. It was not unusual for Devil Anse to sit at the Olive Garden and pay for five or six dinners and gallons of drinks. And he was no fool. He knew they expected it. He called them the *usual suspects*. It was about a fifteen-minute ride from the Bing to the Extended Stay. As I put the car in drive, Sarge turned to White Chocolate and said,

"Not. One. Fucking. Word."

Eight seconds later, White Chocolate said, "*Yo dahhhg n'shit I don't know why you gotsta dis me like that dahhhg n'shit cuz I—*" and

Sarge slapped him so hard the Explorer lurched onto the shoulder with the impact of White Chocolate against the rear passenger side door. He curled up in a ball and whimpered the rest of the way back. Apparently White Chocolate really showed his ass at the Bing. Sarge was as rough-and-tumble as they come, but when he got down to doing business with business*men*, he knew how to comport himself. He knew that even with some bikini-clad bimbo shaking her tits over your schnapps, there were still protocols to be observed. Sarge cleaned up good. He knew the proprieties. And that did not include White Chocolate grabbing dancers' oiled asses, throwing up all over himself, getting kicked out twice, and being let back in by saying *I'm with Sarge*, who by then had become a preferred customer. A regular whale. White Chocolate was lucky the bouncers didn't snap his fingers like week-old bread sticks and toss him in the dumpster, but he was comically unaware of that. Comically unaware that Sarge saved his neck and bought him a clean *Sopranos* T-shirt to replace the puke-covered one he had been wearing. Comically unaware that when Sarge said,

"Not. One. Fucking. Word,"

by God, he meant it. I kept my eyes on the road and my mouth shut. This was a family thing. When I dropped them off under the veranda at the Extended Stay, Sarge said, "Thank you, Doug," again and handed me two hundred bucks.

"C'mon, Sarge," I said. "It was just up the street."

"You're in charge of the clams," he said, and he went inside. And I understood.

That same evening rolling back from the Amboys, I punched *CLAM* into the Dash Bitch just to see what would happen. She valiantly, haplessly, spat out a dozen possible destinations. ExCLAMations Nail Salon, C & L AMusement Company. Five Towns ReCLAMation Services. Jon C LAMb Attorney at Law. No help. We pulled into a McDonald's parking lot and tapped into their Wi-Fi on Crack Baby's laptop. I don't know what he typed into the search engine, but the first hit was PassaiC LAMb of God Tickets. That got the Catholic in me giggling for the next fifteen minutes.

Clam of God, you take
away the sins of the world . . .

It turned out there was a Pathmark not far from the hotel with a good seafood section, and I spent Sarge's two hundred bucks on a bushel of cherrystones, a bushel of top necks, ten pounds of steamers (or piss clams), six pounds of butter, and six cloves of garlic. It was almost nine when we got back to the Extended Stay, and the lot was a'rockin. Jim, Scotty, and Sarge had a pop-up all to themselves, where they entertained five managers from GeoDyne Tech, a handle of Crown Black and clean new glasses on the table between them. Vic with a D, Super Mario, and Triple J were at the grills. The rest of the guys were sitting about on flatbeds and cable spools and some were just getting back from work. Daryn introduced me to two new guys, Frogger and SeanDog, both up from South Florida, but I didn't have time for niceties. I had clams to shuck.

Because of the capricious nature of our manpower, and the constant shuffling of men from one crew to another and from one job to another, all the employees of K.M.C.A. and Sunburst Tower Systems were expected to learn and to perform every task required from load-in to wrap-up. Top hands were supposed to know ground work and vice versa, but that system was as Pollyannaish as it was unenforceable. Workers would naturally gravitate toward skills that suited them; good crew leaders and construction managers would pick up on that and adjust. Angelo Kilfoyle was better at this than anyone, including Brody. And I wished he was in the Jersey market. He could assess this strange brew in a few days and build crews that meshed. It was the same off-site. Everyone had his specialty. And during downtime I was always impressed by what some of these men could do when they were not being tower dogs. Jimmy would work on our personal vehicles. He built hot rods from scratch just for fun, and at any given time he'd have three or four hoods up, making our cars and trucks purr. Daryn was our sound-maven, fidgeting with speakers and amps and wiring, combining iPods with portable CD players and wireless woofers, taking old cassette and eight-track systems and making them viable. Scotty was a gamer and could bring the guys up to speed on the latest developments thereof. Even the Tai Chi guy in the pajamas had an avocation. He was *not* wearing pajamas. Those were his pants. He made his own clothes, and if you had a ripped pair of jeans or a busted zipper, he could fix you right up. Devil Anse baptized him *Homespun*.

Sarge was the master of meat. Vic with a D was the sultan of the side dish. Angie was the reigning king of the crawfish boil, oysters and étouffée, boudin and blood sausage. Brody was the regent of road kill. When we lived in New York, he was coming back from stripping up in Albany and hit a deer on the Major Deegan. Any concern for the crumpled front end of his Jeep fell away as he got out his bowie knife and proceeded to dispatch and then to field dress the stricken animal on the side of the highway. Brody's glass was always three-quarters full. When the state trooper that pulled up inquired, *What the fuck do you think you are doing?* Brody said, "I just want the backstraps. You can give the rest to the orphans." A few hours later Brody was in my kitchen while we slowly roasted the venison. He sat at the table counting out his take from stripping, about five hundred dollars in mistreated singles. "You can help me count," he said, "but I wouldn't put them in your mouth." Once we picked out the gravel and glass from the backstraps, the venison was superb.

And me? I was, am still, and will always be the shaman of shellfish. I set up a small workstation for myself outside a POD and dove in, wielding my clam knife in a blur of precision. I shucked the cherrystones by the dozen, slicing the mussels from the shells, letting the juice dribble into a clean five-gallon bucket at my feet. I garnished them on the half shell with olive oil, Italian bread crumbs, oregano, and minced garlic and then popped them onto the grill. I poured the two quarts of clam broth I had amassed from the bucket into a big stainless steel pot, brought it to a boil, tossed in the top necks whole along with a dozen slim slices of lemon, and let them steam themselves open. The gang was devouring them as quickly as I could prepare them, dipping the salty meat into a small pan of melted butter. The GeoDyne Tech guys, who were mostly from the area and knew their clams, were impressed and told me so.

There was a party going on, but I was not at it yet. Daryn and Cady had been policing the area as per Jim's request, having set up trash cans for the recycling of paper, plastic, and aluminum. They brought me a beer, and I took a break. My hands were raw and sported many stinging tiny cuts. Cady asked about the residue of my labor.

"What do you do with the shells?" he asked.

"Throw them in the garden."

He looked totally perplexed, and I saw the opportunity to assess his gullibility. "Yeah," I said, "you throw 'em in the garden, and in a few weeks you got more clams."

"No shit?" he said.

"No shit."

"I thought they came from the ocean."

Daryn shook his head and smiled, his face saying, *Why not? Go for it.*

"They do," I told Cady. "But you can grow 'em too. Like potatoes. They're clambidextrous."

"Cool," Cady said.

That went beyond gullibility, but he was a likeable kid, so what the hell. He sped off on his skateboard, gliding low to the ground in between moving cars and trucks. I looked over at Frogger and Sean-Dog, who were sipping on a jar of East Tennessee shine. Frogger was scruffy and squat, and SeanDog was tall and broad-shouldered. I did not like the looks of them, but that was not fair. I was trying to take their measure, but you really couldn't do that until you got them on the tower.

"Who are they related to?" I asked Daryn.

"Not sure," he said. "The tall guy Devil Anse met a few months back in some bar. The other guy I have no idea."

I started on the steamers. On Long Island we called them piss clams because when you walked along the beach, right where the flattened sand met the surf line, the pressure from your feet would cause them to shoot water two or three feet in the air. Then you dug like hell. They had soft, off-white shells and a protrusion, or foot, that was totally phallic to look at. The texture was much softer than the quahogs, like a barely poached egg yolk. These I lightly steamed in a shallow pan and drenched with butter and lemon and a little white wine. The booming laughter from the bosses and men from GeoDyne Tech declared the soiree a success. The dogs were indeed on their best behavior. Vic with a D had a nice mix going at a reasonable volume. Outlaw Country with the Highwaymen. *Not Fragile* by B.T.O. And some really dirty Oklahoma blues, with Ray Wylie Hubbard slurring,

Strap them kids in.
Give 'em a little bit of vodka
in a cherry coke.
We're going to Oklahoma

It was almost too quiet. The night clerk, a gorgeous little Puerto Rican woman, even came out to play. The dogs surrounded her, but not threateningly so, and she was loving it.

"She sure is singing a different tune," I told Daryn.

"Jimmy had a talk with the manager. He let the numbers play. We've had forty guys here since mid-February. Comes to over fifty grand so far, not to mention the laundry room and the vending machines. Instant conversion. They ain't gonna' sweat a little racket now and then no more."

My work was done, and I borrowed Super Mario's shower and finally douched the heavy. I called Meg, but there was no answer. I went back into the lot, fresh and oddly fulfilled. The GeoDyne Tech contingent had been slowly but steadily seduced into the fold, and that was good for us, and I had had some small part in it. But something was amiss. I went over to Bo and Gunn, who were sitting with Bigfoot on an aluminum bench that looked suspiciously like the one that used to be at the bus stop a block away.

"Try some of my clams?" I asked, fishing for a compliment.

"I wouldn't eat that shit with Gunn's mouth," Bo said. Nice. Here was a guy who lived on cold canned ravioli, stale saltines, and Natty Lite, questioning my very passion.

"But you'll eat rattlesnake and calf balls," I said.

"Fuckin' A," Bo said.

"If you fry 'em right," Gunn said.

"You are the very definition of a Philistine, Bo," I said. Fuck, they all were. That was part of the fun.

"That's *Mistah* Philistine, to you," Bo said. Bigfoot chuckled, stretched, looked about, and began, *We are all*, and we finished his sentence for him . . .

Bozos on this bus.

"I gotta go see Mrs. Murphy," Bigfoot said. "You guys got the door?"

"We got it," said Gunn, and Bigfoot went into the bushes to pee. I looked at the Extended Stay entrance closest to the party, then back at Bo and Gunn. I shrugged, *What's THAT all about?*

"White Chocolate is under house arrest," said Bo.

"And we're on guard duty," said Gunn.

That's what was missing. That's why it was so calm and quiet. Sarge had forbade White Chocolate to come out of his room, and that was that. The dogs brought him food. The dogs brought him booze. I had not seen or heard the little peckerwood all night, and the groove was substantially more enjoyable, especially for Sarge, who I could see was really relaxed. He had been juggling chainsaws for months, and he always gave the impression he had tectonic plates grinding in his head that were about to give. But now his laughter and ease with the guys was unforced, natural. Ecgberht Schmidt, the champ, the slave driver, the ultimate ball buster, was so positively carefree that to most of us he was unrecognizable.

Then White Chocolate escaped.

The glass doors to the hotel crashed open, and White Chocolate, almost running, headed straight for Sarge. Bigfoot was still shaking it in the shrubbery, but Bo and Gunn dropped their beers and executed a perfect angle of pursuit. Daryn came in from the opposite direction, and the three of them made a wall that White Chocolate would not dare attempt to breach. He wasn't that drunk. He wasn't that stoned. He was on something else, or so out of his mind he had a death wish. He stood on his toes and screamed over the big men's shoulders at Sarge. *Fuckn do me like that dahhhg n'shit fuckn all I done for you dahhhg fuckn kick yo fat fuckn ass n'shit dahhhg fuckn muthafuckn dahhhg muthafucka.* Vic with a D turned the music up high. The GeoDyne guys barely noticed the commotion, but Sarge, feigning calm, registered it all. I could see the veins on his neck pulsate and the tectonic plates in his head grind away. Bigfoot, fly open, threw White Chocolate over his shoulder and carried him back inside the hotel. Bo and Gunn followed. They stuffed him inside his room and told him flatly, *Do yourself a favor. Do not come out that door again. You've been warned.*

He did not come out the door again. He came out the window, taking half of it with him, his leg dragging the ruptured screen across the lot. He was screaming the same *muthafuckn dahhhg shit* all over again. The GeoDyne guys saw it now. And everyone could see Sarge growing bigger in his chair. As his comic unawareness abandoned him, White Chocolate saw it, too. The closer he got to Sarge, the more he glanced from side to side, praying, *Somebody, anybody, stop me.* But the dogs didn't give a fuck anymore. He'd been warned. The die was cast. He stood at the bosses' table rambling, knocking over drinks and plates of food, and that is when Sarge slowly rose and dove headfirst into apoplexy. He placed both of his catcher's mitt hands around White Chocolate's throat and squeezed until his nephew turned blue and went limp.

"Enough," said Devil Anse, and he grabbed Sarge by the wrists. Nobody else would dare. But Sarge hadn't had enough. In his eyes was pure malice and singleness of purpose.

He was going to kill White Chocolate.

Not letting go of his throat, Sarge flopped White Chocolate to the asphalt and straddled his chest. His face went from blue to light gray, and his legs stopped twitching. That's when Devil Anse gestured toward us with his head and said, "Little help?"

It took four of us to pull Sarge off. White Chocolate, coming to, hacking up phlegm and gasping for air as Daryn and I dragged him across the lot, resumed his *muthafuckn this and that* mantra until he once again was deposited in the safety of his room. Back in the lot, what was left of Sarge's better angels embraced him, and he and the dogs laughed it off. *I TOLD him*, Sarge explained, *I TOLD him a thousand times. Didn't I tell him a thousand times?* We all agreed. *Yes, you did.* The GeoDyne guys assured Sarge it was no big deal, but their eyes betrayed them. A blind fool could see they were thinking, *Holy shit, what are we dealing with here?*

Sarge partially salvaged the night by calling in a van he kept on call and invited the GeoDyne guys and all the dogs to the Bing. Two of the GeoDyne managers went along with as many dogs as would fit. There were still enough of us left to get the party back on the rails, and we, saying nothing about the incident, drank and packed leftovers and played music and carried on as always. "Ain't no thing but a chicken wing," Bo said.

By the time my witching hour rolled around, about seven of

us were still hovering. Quietly. Jimmy and Scotty, opting not to go to the Bing (they seldom did go), had gone back to the office hours earlier to plow through paper. The night clerk had taken a definite shine to Daryn. He and The Hayseed Cornpone Fucks had gone acoustic, and it still sucked but not as bad. I had been up for twenty-four hours and would go to work in two if anyone showed up to go to work. I sat by the POD, up to my ankles in discarded clamshells. I tossed them one at a time into a trash can, muttering,

Clam of God, you take away
the sins of the world.
Have mercy on us.

Clack.

Clam of God, you take away
the sins of the world.
Have mercy on us.

Clack.

Brody Kilfoyle was one of the best friends I ever had. He was, I would joke, the first call I'd make if stuck in a Mexican jail. There is nothing I would not do for him, and I thought there was nothing he would not do for me. And my position at that moment was conflicted. I was not Brody's eyes and ears in the field. I was not his spy, though there were dogs who thought otherwise. Neither of us would like that very much. But if he asked me a specific question, I would not hesitate to supply the sometimes-disturbing truth.

What really happened to the '05 Dodge?
Devil Anse used it to knock a citizen off the Palisades
Parkway.

How come it took your crew eight days to complete that
monopole in Ramapo?
Vic.
I got seventeen hundred dollars in toll violations on
the Garden State Parkway. Who the fuck is running
tollbooths?
Everyone without an E-ZPass.
Who doesn't have an E-ZPass?
Everyone.
Are we even supposed to HAVE trucks and trailers on the
Garden State Parkway?
About that . . .

I wanted to call him up, right then, and say exactly what Jimmy
had said. *It's fucked-up up here.* This was not a family thing anymore.
I debated this as the clamshells clacked into the can.

Clam of God, you take away
the sins of the world.
Have mercy on us.

I called Brody but hung up before he could answer.

Clam of God, you take away
the sins of the world.
Have mercy on us.

I called Brody and got his voicemail.

Clam of God, you take away
the sins of the world.
Have mercy on us.

I called Brody and left a message. A long one.

Clam of God, you take away
the sins of the world.
Grant us peace.

Clack.

We were granted peace the next morning when Sarge gave White Chocolate the dog. A cab took him to the bus terminal in Paterson, which was worse than Newark, only smaller. But in this high-stress work-hard-play-hard psychotic petri dish we had created, there could be no lasting peace. SeanDog and Frogger would see to that.

MARCH 25, 2014

The two workers who died after two communications towers collapsed near Blaine (Kansas) Tuesday have been identified, and the Federal Occupational Safety and Health Administration is involved in the investigation.

Pottawatomie County Sheriff Greg Riat has identified the men as 25-year-old Seth Garner of Saint Peters, Missouri, and 38-year-old Martin Powers of Saint Charles, Missouri.

They died Tuesday while working at the 250-foot level of the telecommunication tower that collapsed. Riat said Powers died at the scene and Garner at a local hospital.

13 News has learned the two men had worked for Wireless Horizon of Saint Louis for less than five months. Wireless Horizon is a subcontractor working for the Union Pacific Railroad.

The two men were dismantling an old tower that was right next to a new tower when the accident happened, destroying both towers.

Michael Moon, acting director for OSHA (Occupational Safety and Health Administration), says there were thirteen tower-related fatalities last year and there have been four so far this year.[48]

48. *WIBW*, Topeka, Kansas, "Two Men Identified, Tower Related Fatalities Increasing In Kansas," March 25, 2014. www.wibw.com/home/headlines/Two-Kansas-Towers-Reported-Down-Possible-Injuries-252268221.html?hpt=us_bn9.

CHAPTER NINE MISSISSIPPI
DING-DONG, THE BITCH IS DEAD

I am convinced there is a place in South Central where you do not have to go to school. A place where they just feed you until you are field size and send you out into the world to cut some kind of acceptable swath. You must be this tall to enter society. A place where the teachers took one look, threw up their hands, and went four-wheeling. I am not talking about Los Angeles, but South Central Florida. Southeast of Tampa–St. Pete, due north of Lake Okeechobee, nestled amid the corrugated sea of orange groves is a pocket of poverty, discontent, and disregard, salted with meth labs and rednecks and fruit stands groaning under the weight of rotting oranges and grapefruit. It is a place where the shit-dripping heat and humidity are humbling, without the benefit of ocean breeze, and it is unwaveringly Republican. Due to the company's imbedded Southern sympathies, we seemed to be getting a lot of new blood from down thataway, and few of them, if any, ever panned out. These men were the *Children of Andrew*, the horrific hurricane of 1992 that caused between thirty-two and sixty-four billion dollars in damages to the region, depending on who you ask. Andrew was their godsend. They had decided long ago that picking oranges was beneath them and left that to the army of migrant and principally illegal workers. The reconstruction boom in the wake of Andrew, which created thousands of low-skilled, decent-paying jobs, was pure manna. These men became the sheetrockers and the roofers and the framers and the tree trimmers and landscapers, and after about five years, the boom they thought would last forever began to wane and then finally busted, and when it was over they found themselves flat back on their own heavily potholed jump street. Some of them had a cousin who had a brother-in-law who knew a guy who was working another boom, the big communications build-out up north. Some of them got advanced $455 and a plane ticket, picked up at Newark airport, and deposited in the parking lot of the Extended Stay, Ramsey, New Jersey—as did SeanDog and Frogger, who were, bar none, the two stupidest people I have ever met.

Codifying stupidity is not an exact science; I will admit it is very subjective. But certain truths trump subjectivity. Most of our Children of Andrew could barely read above a third- or fourth-grade level. There were the usual excuses: *My glasses suck. Why do I have to know that?* They could write their names when they absolutely had to. Navigating their pay stubs required both math and reading comprehension, and many Thursday nights were spent explaining to them that *this is your per diem, this is the deduction for your advance, this is your social security.* Their feeble grasp of geography was often an entertaining fissure in their primary education. Clinger was a South Central man in his mid-twenties. Jimmy and I had been sent to do some tower inspections and small repairs north of Medicine Lodge, Kansas. I mobed directly from home, and Jimmy picked up Clinger at MCI. He was a brother-in-law of a cousin of a friend of a worker we had fired months before. And he was green as peas. During the six-hour drive from K.C. to Dodge City, Clinger wanted to know where all the water was. Jimmy tried to point out the Arkansas and Walnut and Cimarron Rivers and the dozen or so man-made lakes, but Clinger was confused. He had meant the ocean. Jimmy explained that there was one ocean about fifteen hundred miles to the east and another ocean about fourteen hundred miles to the west. Clinger could not wrap his head around that, bless his moronic heart. *Bullshit*, he said. In his atlas it was physically impossible for the ocean to be more than a two-hour drive away.

"Where'd you get this guy?" I asked Jimmy later.

"South Florida."

"Is it me, or . . . ?"

"It's not you. Must be something in the crack down there."

We called him Clinger because that is what *he* called the recliner in his motel room. *Hey, it moves*, he said, agape in wonder. It was in that room I had to teach him how to use the phone to dial the front desk. It was also in that room that Jimmy deduced he could not differentiate colors.

"He's colorblind?"

"No," Jimmy said. "He doesn't *know colors.*"

And he didn't. He could tell red and green from his experiences with traffic signals. And sometimes yellow. But ask him to get the blue extension cord from the job trailer, and he'd bring back the orange one. Ask him to grab the gray color-code tape, and he'd bring the blue. Clinger was definitely dangling from the clapper of the bell curve, but the others weren't far behind.

"As long as he can climb and work, I don't care if he thinks Connecticut is in Canada and the Stars and Stripes are purple and pink," Sarge said.

To me, the most egregious of their misfortune was their total lack of historical perspective. You can throw a dart at the map of this country, and no matter where it hits, something important or at least memorable has taken place there at some time. And in our travels we would find ourselves in the most incredible places, often now reduced to dust or a decaying highway marker. Pottawatomie Creek, Kansas: John Brown and his family took broadswords and went Biblical on some Missouri ruffians and basically kick-started the Civil War. Narrows, Georgia: Wow, Ty Cobb was born here. Eldon, Iowa: the American Gothic House. Grant Wood stood right here on the curb and painted that. Paterson, New Jersey: Who's that a statue of? Lou Costello? *Damn.* Titusville, Pennsylvania: This is the site of the country's very first operating oil well. *This* is the O.K. Corral. *This* bank was robbed by John Dillinger. *This* is where Custer got his ears bored out. *This* is the Continental Divide. *This* is where Tammany sold Manhattan, and *THIS* is the goddamned Plymouth Rock. You can't blame this on history teachers. Some of this had to be taught to them at some point. Their accumulative knowledge of history seemed to consist of what happened on *TMZ* or *American Idol* the night before. They did not want to see or to know these places. They wanted to go to Barbie's Tuck-A-Buck. And they would tell you, almost with a touch of inept pride, that

all THAT happened BEFORE I was born

as if anything prior to their birth like Pearl Harbor or the Battle of Hastings or the Johnstown Flood or the Emancipation Proclamation was inconsequential.

Listen up, children. Today is August 19, 2007 A.C. (after Clinger). Nothing really happened before HE was born, but NOW you have to pay attention.

Not surprisingly, there were subjects in which they could be considered experts, like how to cheat the U.A. and what were the best new apps on their cell phones and the going price of the hottest footwear. If these guys were required to fill out their applications at the main office on Navaho Road in Nashville, half of them would not have made it past Katt, Brody's wife and the head of human resources. Being married to Brody and having two toddlers, her bullshit detectors were as finely calibrated as my father's, and she could smell a mistake before it walked in the door. But due once again to expedience, a lot of the new hires were going straight from brother-in-law to market. And that was one of the oddly magnificent things about K.M.C.A. and any philosophy K.M.C.A. might have inadvertently chartered: K.M.C.A. *wanted* you to not be a fuck-up. K.M.C.A. *wanted* to take you from the crap that was your life and offer you something better. K.M.C.A. *wanted* you to succeed. K.M.C.A. *wanted* you to grow and learn and make more money and make them more money. K.M.C.A.'s glass, an extension of Brody's, was always that three-quarters full. K.M.C.A. *wanted to believe in you*. And for the most part, it worked.

The majority of tower hands and the majority of K.M.C.A., despite their peculiarities, are your basic blue-collar workers with a measurable scale (however fluctuating) of ethics and morals and common sense and the ability to balance the craziness of the job and the road against the pressures of a neglected home life and love life and the world that *does still exist* outside their own eccentric bubbles. The reason K.M.C.A. got so much work in the first place WAS these men and the timeliness and the quality of their work. *The job so nice we did it twice*, Devil Anse would say, referring to the times K.M.C.A. would whittle away at their own profit margin to appease the mercurial impulses of the carrier (or agent thereof), sending crews back to sites that were perfectly implemented just to reroute a jumper or beef-up some coaxial support. One night Daryn and I rolled to a site with a truck and trailer, a good six hours of overtime, to shift—*not replace, but SHIFT ½₂ of an inch*—a ⅜-inch washer on a ground lug at 140 feet because the GeoDyne inspector said it looked "off-center." Piddly-ass things like that made Sarge's plates grind and Jimmy and Scotty just shake their heavy heads in wonderment. But K.M.C.A. did deliver. And our core men were right and tight. And in being that, most unfortunately, these core men were *not* at all memorable. That is the residue, the fetid byproduct of the advent and popularity of reality television. You don't remember the crab wrangler who is a stable *do-my-damn-job-and-go-home* family man. You don't remember the articulate logger on break from college earning money to pay for his Comparative Anatomy books. You remember the asshole. You *want* the asshole. You want the asshole with the bleeped-out sound bites. The asshole expounding ad nauseam the ever-present and totally contrived DANGER. The asshole who is the fulcrum of the internal crew CONFLICT derived from whole cloth. Even before the Tower Dog Special aired, just based on the twenty-second promo, the folks at Peacock Productions had offers from four cable networks to pick up the show as a reality series. One of the bidders was the History Channel. At a meeting in their offices two days after the air date, one of their development gurus actually clapped his hands and said, in a preposterous Australian accent, "So I'm guessing we 'av a good supply a' toothless bearded rednecks, 'av we?"

We did not.

Our conflicts were real. Our danger was real. Our work was very real and often so flawlessly executed it went completely unnoticed. As did men like Jess Pulaski, the laconic redheaded Tennessean who carried every tool he owned up the tower so he would not have to impose on the groundhogs to send them up. A man who, when burdened with a slap-dick crew, would just do it all by himself and not utter a word of complaint. After work he'd drink one beer, check the fluids in his rig, go back to his room, call his stepdaughter, and update his close-out packages and JSAs. As did men like Hangman, who baptized me on my first tower, practically a kid himself at the time, who was all about getting it done quickly, efficiently, without drama or hyperbole. If he felt, which he often did, that communication with the groundhog at the site house was hampering his progress, Hangman would just hurl the radio off the tower and not deal with it anymore. As did new dogs like young Cady, who was fast becoming a capable and competent top hand, the kind of guy who could advance to crew leader in a matter of months. When I talk to the veterans of the Jersey Market, they never recall the Pulaskis and Hangmans and Cadys without prompting. But they all remember the White Chocolates, the SeanDogs, and the Froggers. They remember the guys who never had passed or perhaps even taken the unwritten test that said *this is how we generally behave when with people other than ourselves. People who do not know us. Just a few tips. A primer, if you will, that says keep your mouth shut and your head down, and maybe you could make a nice living doing this. This could actually be a career. Occasionally fucking just listen.* Instead they landed like retarded geese on ice, flapping and sliding, drawing unwanted attention to us all.

Their second night in market, Frogger and SeanDog, who might have formed some half-assed alliance due to shared geography, had teamed up and managed to alienate almost every person they came into contact with, tower dog or no. They had co-opted a pop-up and were far drunker than they should have been at that hour when Bigfoot, Daryn, and I pulled in. A few of the other dogs were sitting about, and I could tell by their faces they were trying to sort it out. Mo and Ron, also from South Central, kept a wary distance as they detailed their truck cab despite SeanDog's pleas to *fuck that shit and come have a seat, homies.* Cady sipped a Mountain Dew and flipped the skateboard at his feet. The Rheingold guy had stopped by again, and even he was reticent. Bigfoot took one look, said *I don't want any part of this*, and went inside. Daryn went to visit the desk clerk, who had refused to come out because the Children of Andrew were on her like sweat. I sat down and grabbed a beer and was told, *Hey, that ain't for everybody.* And that set the tone.

In their defense, all dogs are fish out of water; we find ourselves in strange new lands with strange new people from market to market, and we're not always caught up on the local playbook. And being the new guy in a loose fraternity of compadres cannot be easy. A little bravado is expected. A *little*. A little trying to be part of something bigger. I listened quietly and smiled and nodded as SeanDog and Frogger blasted holes in their own rowboat.

*. . . This job ain't shit compared to hanging red steel in
Miami . . . My cousin said this job was hard . . . Ain't shit
. . . We got to get to the city . . . Who's got wheels? . . . Who
can take us to the city, dog? . . . You the goddamn Yankee?
. . . Fuck it sucks up here . . . Just kiddin, bro . . . Where'd
that fine-ass spic go? . . . Who's the faggot in the pajamas?
. . . Where's all the pussy at? . . . Just kiddin, homes . . .*

I tried to explain to them in no uncertain terms that I was not their *bro* and have never been anybody's *homes* and that even if I *knew* where the pussy was at, I wouldn't give them directions. I told them to punch *gonorrhea* into their Garmin. They responded with, *Ooooh, he's sensitive.* They had been in market for less than twenty-four hours. I leaned in to Cady and said, *I give them a week.* Frogger would not last a week. He would not even make it until dawn.

If someone told me I would be using the names Tammany (sachem of the Lenape) and Ansobert Schmidt in the same sentence, I would lay heavy odds against it. But both men had a keen eye for the placement or misplacement of things, specifically infrastructure. "Whoever built New Jersey was an asshole," Devil Anse said, and I think Tammany would have grunted and agreed. When it came to certain idiosyncrasies in New Jersey's road system, Devil Anse was right. The stretch of Rte. 17 we negotiated daily served as the perfect microcosm for Devil Anse's contention. It was just one really fucked-up piece of engineering. Rte. 17 (in a route system that numbers from 3 to 440) is a ribbon of asphalt heading north and south, two lanes each way separated by a three-foot-tall concrete divider. At every exit was a vehicular abortion known only to New Jerseyites as jughandles.

A jughandle is a type of ramp or slip road that changes the way traffic turns left at an at-grade intersection. Instead of a standard left turn being made from the left lane, left-turning traffic uses a ramp on the right side of the road. In a standard forward jughandle or near-side jughandle, the ramp leaves before the intersection, and left-turning traffic turns left off it rather than the through road. Right turns are also made using the jughandle. In a reverse jughandle or far-side jughandle, the ramp leaves after the intersection, and left-turning traffic loops around to the right and merges with the crossroad before the intersection. The jughandle is also known as a Jersey Left due to its association with the U.S. state of New Jersey, though this term is also locally used for an abrupt left at the beginning of a green light cycle.

Ergo, a Jersey Left is a left you cannot make.

All of Tony Bill's horses and Tony Bill's men could not make that left turn. And you do not have to be a Child of Andrew to not understand this. But practically applied to us visiting dilettantes, it usually meant *you can't get there from here.* Between the one-way two-lane highways and the concrete divider and the jughandles, we were constantly going around and around and around. The Dash Bitch also could not make heads or tails of it and just said *screw this* and turned herself off after recalculating about a hundred times.

On our side of Rte. 17, about seventy yards past Lake Scum, was a Tiger Mart that you could walk to for soda or cigarettes or coffee or chips. You could walk there and back in about three minutes. But if you *drove*, it would take twenty minutes. If you drove, in order to get back you had to go north on Rte. 17 another 2.5 miles, exit right to the east, go around jughandle 1 to the south then west, take jughandle 2 north then south, head five miles back down Rte. 17 past the Extended Stay on your left, exit right at the first Upper Saddle River exit, take jughandle 3 to the west then east, take jughandle

4 to the south then north, and head two and a half miles back to the Extended Stay. Frogger, in a short-lived moment of lucidity, thought this to be ridiculous. And Frogger wanted beer. But the Tiger Mart did not sell beer, and none of the new friends he made were willing to drive the ridiculous route to get him some. There was a Hess station across Rte. 17 that did sell beer, but nobody in his right mind would try to cross it. The speed limit was fifty-five, but they all did seventy, and this was no country lane. It was a vital vein to the suburbs and shopping centers, and traffic was always thick and fast.

Vic with a D and Crack Baby had come out of their rooms freshly showered and turned up one of Vic's mix-tapes. Crack Baby, though seemingly too young for it, had a great appreciation of the album rock of the '60s and '70s and was rifling through Vic's impressive selection of Led Zep, The Beatles, E.L.P., and the Nitty Gritty Dirt Band. Right in the middle of "Buy for Me the Rain" we heard the screeching of many tires, honking, and cursing.

Frogger was playing Frogger.

As SeanDog stood on the side of the road egging him on with rebel yells, spilling his drink in the air, Frogger was dodging cars, trucks, and buses on his way to get his beer at the Hess station. He made it back alive somehow, cradling the six 40 Dogs of Colt .45 malt liquor like Diesel Riggins, causing as much commotion on the way back as he did on the trek over. He and SeanDog whooped, exchanged sloppy high fives, returned to the pop-up, looked over at us, and shouted, *Y'all're a bunch a pussies.*

Secret Squirrel and Dickless Tracy showed up about twenty minutes later, inquired as to the disturbance, were told *it wasn't us, sorry,* did not believe that at all, and went on their way. Pulling out in their cruiser they shot a *we have badges* look at Rheingold Guy, who waved and asked how they were doing that fine evening. Frogger would play Frogger not once, not twice, but three times, heading back for beer and cigarettes, either not thinking far enough ahead to buy it all at once or trying to impress us with his nonchalance regarding self-preservation. We were all thankful when at eleven o'clock the Hess station closed. But the most codifiable stupidity of the Children of Andrew was still to come because on one of his hops across Rte. 17, Frogger had looked south toward the New York City skyline. He had seen what he needed. He had seen the promised land. And nothing that night would keep him from it.

I was pissed now. I was so pissed that when Brody returned my call from the night before, I did not even answer the phone. I did not want to have to tell him that I saw a bad moon rising. That though we were a little bit on the cops' radar before, we were a lot on it now. I sat with Crack Baby and stared at the ground while Vic with a D's stereo was on to "Karn Evil 9."

Welcome back, my friends
To the show that never ends.
We're so glad you could attend.
Come inside, come inside.

That was way too on the nose, and I asked Crack Baby to pick another CD. He chose Grand Funk Railroad, but I could not enjoy it because even though SeanDog and Frogger couldn't quite comprehend that I was not listening or responding to them in any way, they persisted in asking me to take them to the city about once every fifteen seconds. I knew why they wanted to go to the city hours ago, and it wasn't to see the frigging Guggenheim. They wanted to score something other than beer or a little pot, and they knew cities were a good place to do that. Frogger went inside, and I could hear him banging on doors. SeanDog passed out in his chair, mouth open to the moths. I had played Den Mother to lots of dogs over the years. Shouldered them off barstools and into the backs of trucks. Escorted a sleepwalking Bo back into his room and slid him under the covers, dodging roundhouse punches left and right. Recovered their misplaced cell phones and car keys from wherever they had dropped them. Normally I would have woken up SeanDog and pointed him toward his room, but not tonight. I crawled into the back of the Explorer. He could eat bugs.

When my father died, he left me two things: one of his Green Beret Ka-Bars, because he knew I loved to fiddle with his knives, and a two-dollar bill minted in 1976. I don't know what he was thinking at the time, because his brain was porous with cancer, but it didn't matter because it came from Dad and I associated some importance to it. I kept the two-dollar bill folded and tucked into my wallet, and I kept the big knife in its sheath tucked between the driver's seat and the center console of the Explorer. And when I tried to sleep at night I always took off my shoes, rolled down the windows, laid on my stomach, and placed one hand on the hilt of that knife. It was a ritual—of which I have many—that served no purpose except for being a ritual. Like kissing the dog tag. Like carrying around a filthy rubber duck in my front right pocket.

The witching hour came early. At about 3:30 A.M., I awakened to the sound of sirens. Police and fireman and perhaps an ambulance, from what I could tell. By the time I shook off enough sleep to extract myself from the Explorer, they had passed south down Rte. 17. I stood there, relieved, unaware I had my knife in my hand, and felt guilty I had left SeanDog slobbering in his chair. I would throw a blanket on him, but when I got to the pop-up, he was gone. I dragged myself back into the Explorer hoping to get at least another hour in before muster. I felt something warm and squishy between my toes. I had stepped on something and was bleeding but not enough to keep me from trying to sleep.

Next thing I heard was GODDAMNIT.

Devil Anse said *goddamnit* about a thousand times a day, and we had learned in time to gauge the urgency of the expletive. This was not an *Oh, hell, I spilled my coffee* level of *goddamnit* but a ten with a bullet on the Ansobert scale of *goddamnit*s. He was genuinely pissed, and when I asked why, he pointed to two work trucks about twenty yards away with their side windows blown out and doors wide open. The guys had been told by Jimmy and Sarge and Devil Anse every damn day to take their Garmins into their rooms at night and not leave them stuck to their windshields, but they never listened. Devil Anse didn't care about the Garmins. He wanted to go to work. It was 4:45 A.M., and he was already behind schedule. Now he had to deal with the cops and the insurance adjusters and

GODDAMNIT DELANEY YOU SLEPT THROUGH THIS?!

I had. I also couldn't offer up any information as to the time of the crime because I did not even look at the trucks after the sirens woke me up, even though I had obviously traipsed through the shattered glass. The blood on my sock had dried, and it felt like I was wearing bacon. The dogs were assembling, and I was astonished to see SeanDog dressed in a clean white T-shirt, jeans, and work boots, fresh as a daisy and ready to go to work. I saw Sarge steer him toward Daryn and thought *Oh, Christ, PLEASE not on my crew.* We unhitched the stricken vehicles, and the daily cavalcade of K.M.C.A. began to trundle out of the lot toward parts yet unknown to most of us. Sarge stared at the damage, his brow crinkling, rearranging the day he had planned in his head so it could still work without two of our trucks. Devil Anse was kicking at the glass repeating *Goddamnit I ain't got time for this shit goddamnit* as the cops pulled in.

"That was fast," Sarge said.

"I didn't call the cops," Devil Anse said. "Delaney, did you call the cops?"

I was just getting into the truck cab behind SeanDog, who had taken shotgun, and that was something new meat *always* knew enough never to do unless told to. It was a minor place of honor among dogs. Shotgun was something you earned.

"Not me," I said.

These were not Ramsey cops. These were state troopers. And after a few words with Sarge, they were off again, leaving him a page from one of their notepads. Sarge looked at it and then closed his eyes. "Unbelievable," he said. He looked around the lot assessing manpower and chose me.

"You stay here," he said. "Call the cops. Get a glazier out here. Devil Anse, give him the number for the insurance. I gotta go." And he went. Devil Anse went. They all went. And I stood there crunching my toes in my flaking sock thinking *Glory be, I am Vic for a day.* Super Mario left me his room key, and I made a pot of coffee and sat outside in the Dale Earnhardt chair, reveling in the calm, enjoying the exceptional morning where I could actually sit and drink a whole cup of coffee before mobing. There were cattle egrets stepping tentatively about the shallows of Lake Scum, looking like little yellow-combed storks. The Ramsey cops, two different ones, showed up around seven. In a statement that sounded rehearsed, one cop said that *it is a documented fact that currently the Garmin is the most pilfered item in the state of New Jersey.* This was somehow, of course, our fault. We should have known better. I took the paperwork and called the insurance company and left a message. Business hours. This was great. I could count on not getting *that* call back until nine at the earliest. I had been averaging less than five hours of sleep a night, so maybe I could squeeze in a few desperate winks. But five minutes later I got a call. It was not the insurance guy. It was Jungle Boy.

"What the fuck, Delaniac," he said.

"I know, I know," I said. "I'm waiting on the adjuster now."

"What adjuster?"

"*What?*"

"Is he alive?"

Brody and I could often read each other's minds but only when we were in sight of each other, and on this one, we were not on the same page. Hell, we weren't even in the same library.

"Is the cocksucker alive?"

"Who? I'm waiting on the insurance guy. They got two Garmins and—"

"They left the Garmins on the windshield!?!"

"Yeah. And Sarge said—"

"Fuck that. Is that goddamned waterhead alive because if he is *I am going to come up there and kill him myself.*"

This call wasn't about Garmins. This was about Frogger. He had seen the promised land down Rte. 17. He had seen the promised land, and nothing was going to keep him from it except the bakery truck that ran over him at about 3:15 A.M. as he was attempting to cross the concrete median down by the first Upper Saddle River jughandle. He was in the hospital. He was alive. Bloodied but unapologetic. Bowed but unwizened. Sarge was en route to the hospital but running a little late because he had already stopped and bought Frogger a bus ticket back to South Central Florida. Frogger had been in market thirty-two hours before getting the dog, a company record that stands to this day.

JUNE 17, 2014

A man fell to his death while doing repair work on a downtown telephone tower Tuesday afternoon, according to a San Angelo Police Department news release.

Tom Green County Precinct 4 Justice of the Peace Eddie Howard pronounced the man dead just after 1:50 P.M. Tuesday, the news release states.

The man was a contractor with Richardson-based MTSI Inc.

The man fell about 140 feet from the tower, at 378 South Chadbourne Street. He experienced issues with his safety equipment before falling, the news release states.

The San Angelo Fire Department and Police Department responded to the scene.

In the past five years Occupational Safety and Health Administration has cited MTSI Inc. for three serious safety violations. OSHA fined the company $600 regarding a head-protection violation in 2010, and $1,500 regarding cranes and derricks at the same time. It was fined $2,520 regarding aerial lifts in 2012, according to OSHA reports.

The San Angelo Police Department has released the identity of the man, who fell to his death from a downtown telephone tower, as Cody Freeman, 28, of The Colony, following notification of next of kin.[49]

---

49. *San Angelo Standard-Times*, "Man Who Fell to His Death from Telephone Tower Identified," June 17, 2014. archive.gosanangelo.com/news/update-man-who-fell-to-his-death-from-telephone-tower-identified-ep-503239426-354770271.html.

CHAPTER TEN MISSISSIPPI
MOM, I THINK I WAS IN HEAVEN TONIGHT

It was 5:00 A.M. The sun not up. Maureen McWilliams sat wrapped in her faithful blue robe at the kitchen table, waiting. The robe was played out, and Ted had tried to buy her a new one a dozen times but this one was just perfect. It fit her when she was twenty-four, and it fit her now. It grew into her as she grew into it. She was not startled when she heard Jonny's boots stamping on the welcome mat outside the kitchen door. He had been out all night. She had been up all night. That is what mothers do in Iowa. She could hear her husband Ted getting ready for work in the upstairs bathroom. He was moving slower these days, MS having tightened its grip. They both knew there was too good a chance for a wheelchair in Ted's future, a premature set of wheels he was privately dreading but would do his best in the meantime to ignore. It was late spring, and even though the mercury would hit ninety that day in Cumming, just a bit south and west of Des Moines, the mornings still held a welcome chill, and you could open the windows and get a good dose of brome-scented unrefrigerated air for a while. They had barely had a spring. It seemed to have gone from bone-numbing plains winter to pasture-crackling plains summer without even letting you get your seeds in order. Jonny ran his hand through his black hair, just shy of a crew cut, kissed his mom on the top of her head, and opened the fridge.

"Don't tell me you've been up all night," he said.

"Don't be silly," Mom said. "You come and go so much. Meghan is trying to get a hold of you."

His sister Meghan needed his shoe size. He had a birthday coming up soon. In three days he would be twenty. He and Meghan were as close in life as they were in age. She had his same thick, dark black hair, only longer, and was trying to put herself through beautician's school. The McWilliams always found that ironic because Meghan was a woman who could easily grace the cover of any fashion magazine. But she didn't think like that. She was going to make others beautiful. She had the six-month-old, Bella, but she always found time for Jonny. When she moved into the new condo, she and Jonny got a bunch of pizza and beer and painted the whole place in a weekend. That was a memory that would have to carry her a long way.

Jonny leaned back against the fridge and drank straight out of the milk container.

"She wants my shoe size," he said.

Neither Jon nor Meghan *had* to go to work. In a time when there was still a lingering middle class in this country, the McWilliams were at the upper tier of it. Ted had done okay, thank you, but that did not matter because when you come from Cumming, Iowa, you worked, and that was that. If you asked Ted McWilliams where he was from, he would not say Des Moines.

"She's gonna get me the new Jordans," Jonny said, grinning and bumping his eyebrows. "It's a *surprise*," he said.

"And how did you get so smart all of a sudden?" Maureen said. Having already done well at Hanawalt Elementary School, Merrill Middle School, the Iowa Christian Academy, Winterset High School, and his first three semesters at Des Moines Community

College, the kid was not suddenly smart but damn smart, and that is why Maureen could not understand why he didn't get a job (even if it was just for the summer, like the one he had now) where he could use his brain instead of his brawn. But try to tell a nineteen-year-old anything. He could make as much money behind the counter at the QuikTrip as he could climbing these towers. He could run a register in his sleep. And he could do it at ground level.

"He's just testing himself," Ted had told her, when Jonny announced he would be going to work for Deters Tower Service, Inc. of Des Moines. "You wouldn't understand."

"I do understand, but I don't have to like it."

"No you do not," said Ted. But he didn't like it all that much either.

"Teenagers' brains are . . . *different*," Maureen said.

"That they are," said Ted.

Jonny could not feel the absolute joy and relief emanating from his mother as she gently touched the sleeve of his white thermal work shirt. He was home. He was safe. He had been doing this job for almost three months, but for her every day was his first. He could not see the joy and relief, but he could see in her half smile the worry because he had seen it about a million times.

"I'm *fine*, Mom," he said.

No amount of assurance could ever stem the worry. She had borne the worry when he discovered the knife drawer and when he fell off the monkey bars and when he choked on that damned Super-Ball, and when he first borrowed the car and went white-water kayaking and took his flying lessons. *How he loved to fly.*

She could not acclimate to it. What mother could? And now this . . . this lunacy.

"Was it chilly up there?" she said.

It was a little after 5:00 A.M., and the sun, unobstructed by the wrinkled sheets of pastureland to the east, was coming on murky orange and strong, awakening the wind. Ted came into the kitchen and poured himself a cup of coffee, and Maureen said, *I'll get that,* and Meghan was coming by later with the baby, and Jonny had a birthday coming soon, and everything was as close as it could be to normal.

I had not fallen asleep in the Godfather's Dale Earnhardt chair after I had talked to Brody and learned Frogger went *kersplat* down by Upper Saddle River. I had passed out from exhaustion. The first five or six times my phone rang, I just let it. The insurance estimator had come and gone and the glaziers were in the lot working away and I wasn't going to do a damn thing until somebody, anybody, came to get me and put me to work. I was on the clock. Screw them all. At about ten thirty I finally did pick up the phone, and Laurel said the same thing Brody said:

"What the fuck?"

"Hi, Digger," I said, tower dogs not being the only people with nicknames.

"What *theeeee* fuck?" she said.

"I'm fine, thanks. And you?"

You raise five boys, you can cram anyone into a corner. I was doomed and I knew it and I would just have to ride it out.

"You're *five minutes* from me for *shitknowshow* long and you don't even bother to call?"

"That's because . . ."

I had no because. I was still half asleep but alert enough to wonder how the hell she knew I was in New Jersey.

"That's because," she said, "YOU are an idiot."

"I'm sorry, Dig. I was going to call you the first chance I got and—"

"I don't have time for this," she said. "I got tennis. I'll pick you up at the Extended Stay."

"Pick me up? Laurel, I'm at work. Why would you want to pick me up?"

"Because I have your wallet, you idiot."

It was a morning for odd conversations. The upshot is that whoever stole the Garmins reached into the Explorer and slid my wallet right out of my back pocket. I'd like to think that if I had been aware of this, I would have yanked my dad's Ka-Bar out of the sheath and made a real bloody mess of it. Instead I just slept. A motorist found the wallet flopping on the shoulder of the road up by Pathmark. Inside, all that was left was my license and a card from the front desk of the Extended Stay. And a phone number. Laurel's phone number. She had met the motorist at Pathmark and got the wallet for me, a kindness for which I knew I would pay dearly in about ten minutes or so. Laurel didn't feign mad when she was mad. She just did mad. And when she pulled up underneath the canopy at the Extended Stay, she said *I don't get you. What am I? Chopped liver? You don't even fucking call?* Her cheeks were red.

"You look great, Dig," I said, and she did, but she wasn't going to let me off the hook that easily. "Get in the car," she said, and what else could I do? We went to the Starbucks across the street from Grady's in downtown Ramsey. We sat outside as the off-peak commuter trains came and went. There didn't seem to be any of my former ilk here, and I was thankful for that. People were there to get coffee and catch a train, not to make an impression.

"So, Dougie, I thought you quit this shit," Laurel said, absent-mindedly curling her hair with her finger. She'd been doing that since she was five. She was also the only person left in the world that called me Dougie. I could only shrug a response. She and I had so many elongated and in-depth discussions about my aspirations and failures that I really had nothing to add to them.

"So how's the baby?" she said.

"Just great," I said, but I wanted to say, *How the fuck should I know? He's there and I am here and right now he could be crawling across a fire ant hill in his diapers and not only would I not know about it but what could I possibly do about it?* Nobody at work ever asked me, *How's the baby?* And it hurt to hear it. Because I truly did not know and felt ashamed.

"Just great," I repeated. "How's Kevin and the kids?"

"Don't ask," she said, and she handed me my wallet. They took about two hundred and sixty dollars in cash, all I had, my ATM card, and even my dad's two-dollar bill. Despite her digging at me, which is how she got her moniker, we sat in the comfortable silence of old friends.

"We're going out to the house on Shelter Island for Labor Day. Just Kev, me, and the little ones. Kev says you're welcome to come out if you can."

"If I can."

"I suppose you need money," she said, snapping open her purse.

And there it was. My Digger. Cutting out the bullshit and getting right to the meat of it. Of course I needed money. My ass was just burgled. I couldn't even pay for the coffee. But I wasn't going to tell her that. I had a plan in mind. Besides, I already owed her lots of money. A few years back in 2001, during one of my stunningly naive and optimistic phases, Fallon and I, tired of being encouraged to death, decided we would actually produce our own independent movie. We had a good script, the spec that got me into that business, a well-known director attached, and even some name talent. It was a small budget, about two million. But we needed money to go after money, and Laurel and Kevin invested a chunk. Things went great at first. We shot a trailer (starring Brody Kilfoyle, of all people, who—soap-opera class aside—was a pretty good actor), and I went on the road looking for funds. By late summer of 2001, I had legitimate letters of intent for almost half the budget. Then nineteen affiliates of Al-Qaeda took four unscheduled flights into infamy, and my project died on the vine. Laurel and Kevin never once asked for that money back. But I knew Digger was at the very least disappointed in me, and that hurt as much as the whole thing going south. Like K.M.C.A., she wanted me not to be a fuck-up. They had a lot of money, but it was not play money. She and Kevin both came from modest means and respected it. She grew up in the same nine-hundred-square-foot Levitt house that I did, the one with the unfinished attic, without air-conditioning, central heat, or a sewer hookup. And despite the significance of our past, it was hard to know that in their eyes I would always be Dougie the bad investment.

"I'm okay," I told her. "But thanks anyway."

I was full of shit, and she gave me that *I know you are full of shit* look and snapped her purse shut. She was still enough of a friend to let me dwell in my pretense of dignity. She dropped me off at the Extended Stay, yelled *Shelter Island! Labor Day!* out the window, and drove off to pick up this kid, drop off that kid, and hopefully make her tennis lessons.

I went into Super Mario's room and grabbed his whole Crown Royal bag of change, left a note for him, and then dropped the keys to the two trucks at the front desk and left a note for Sarge. No sense in calling. They would call if they wanted me. I called my brother Robert in Levittown and told him what happened and asked if he could bail me out and he said come on over and I headed toward the George Washington Bridge. On the way I listened to 1010 WINS. *You give us twenty-two minutes. We'll give you the world.* By the time I got across the GWB, through the Cross-Bronx and over the Throggs Neck on to Long Island, I had given them twenty-two minutes six times and profoundly pissed off the toll collectors at the crossings with Super Mario's quarters, dimes, and nickels. The Cross-Bronx and the LIE were the nightmare they always were, but I had no choice but to take them. It was like waiting an hour in the dentist's office knowing things were just gonna get worse. The radio told me that an NBA ref was getting indicted for some gambling scheme, that it was the anniversary of Martin Luther King's *I Have a Dream* speech, and that Don Imus's tongue was digging him an even deeper hole in a world that was just waiting around for something to be virtuously offended by.

Ted McWilliams was listening to his car radio as he drove the eighteen miles down I-35 from Des Moines to his home in Cumming, Iowa. News Radio 1040 WHO reported that it was eighty-four degrees and rising with six-mile-per-hour winds out of the north by northeast and not a cloud in the sky. And that three men working for Deters Tower Service, Inc. had fallen to their deaths in Oakland, Iowa.

"They didn't mention names," Ted said. "I pulled off the interstate and I called my daughter at home and asked her to call the company office and see if she could find out anything. And they . . . the company office was not releasing any information."

Maureen McWilliams was in her car but not listening to the radio. She had gone to see Meghan and Bella in their freshly painted condo, but they had gotten their wires crossed and kept missing each other so she just headed on back. "And then I was driving home," she said. "And we live way out on a country road. All prairie, beautiful rolling hills. And I see Ted behind me in the car—which has never happened in my life."

The office at Deters Tower Service, Inc. was not giving out information, but 106 miles to the west, almost a straight shot out I-80, at the base of a 1,450-foot guyed-wire tower that had now become a crime scene, people were talking. Sheriff Danker was saying, "They'd been up the tower earlier in the day, they'd come down for lunch, and they were on their way up the tower again when this occurred." The winch operator was saying that the crew was going up, the line went slack, and he looked up and noticed them falling.

They were changing light bulbs.

They were into the second day of replacing flash tubes, according to Bill Hayes, chief engineer for Iowa Public Television, which owns the tower. Hayes estimated the men fell from between eleven hundred to twelve hundred feet.

That is a hummingbird's heartbeat shy of Ten Mississippi.

"And in the driveway Ted says, 'There's been an accident. Three guys are down, and I think it's Jonny's team,'" Maureen said. Then her brain splashed hope in her face. *They use more than three guys, don't they? The other night he said there were five of them. Did he say five? Or four? They wouldn't send him up that high. He's so new. They wouldn't put him up that high.*

"I'm calling the company and I'm calling the sheriff's office, and to this day they have not returned my calls," Ted said. "Then, later that evening, two tower workers came to the house and told us one of them was Jonny."[50]

"He was going up the tower," Meghan said. "And he told me his shoe size and said he would call when he got back down to come *hang out . . . see the baby . . . you know.*"

I pulled off the Wantagh State Parkway, still a beautiful road, sunk deeply into the land, laced with good thick woods and huge glacial boulders, the bridges low and built with heavy stones so you could not see the suburban sprawl that surrounded you. No trucks allowed. I stopped at a bank on Hempstead Turnpike and dumped all of Super Mario's coins on the counter. They ran it through their coin machine, for a fee, and handed me 134 bucks. I gave the teller back two singles and asked if she could find me a two-dollar bill, preferably minted in 1976. She did, to her own surprise, and I folded it neatly and slid it into my wallet. I would try to forget that the original was ever stolen, and, many years from now when I gave it to my son, I would tell him that my father had given it to me. And I'd throw in the Ka-Bar for the hell of it.

It was a little after 5:00 A.M. The sun just up. And Maureen McWilliams touched her only son's sleeve and said, *Was it chilly up there?* and Jonny looked up at her and said,

---

50. Ted and Maureen McWilliams's dialogue taken from NBC *Dateline* field interview sessions in Kansas City, 2008. See www.nbcnews.com/video/dateline/25738683/#25738683.

It's so beautiful, it's so exciting, it's so
exhilarating. I think I, really, you know . . .
I love this job.

He had been up fourteen hundred feet that night, and below he could see the smoky glow of sleeping cities from Illinois to Nebraska, from Minnesota to Missouri. Above he could see all the stars and all the planets rotating around the center of the universe, which was, at that moment, him and him alone. He had climbed higher than the tallest spires of all the cathedrals on earth.

And then he said

Mom, I, I think I was in heaven tonight.

JULY 2, 2014

The worker killed while working on a Harrison County cell phone tower has been identified as 28-year-old Joel Metz, a father of four. A cable somehow decapitated Metz and left his body suspended from the tower on Waits Road off Kentucky 36 Wednesday afternoon. A preliminary autopsy report says Metz died from blunt force injuries and his death is an accident. Harrison County Sheriff Bruce Hampton says Metz worked for Fortune Wireless out of Indianapolis, which is near where Metz lived. The company was back out at the scene Thursday morning trying to figure out what caused the cable to loosen, leading to Metz's death. Authorities first believed the cable had broken, but after investigating, the sheriff says the cable was still in one piece. The three other workers at the tower reported hearing a loud pop before realizing what happened, according to the sheriff.

"Not too many people are used to seeing that. And it's a guy that you've been working next to," said Sheriff Hampton. "And it could have been him. Nobody else was injured. And they are just very much beside themselves." Sheriff Hampton says the men were taking down an old boom and bringing up a new one when a cable broke, decapitating one worker and ripping off his right arm. "It got within two feet of where it was going and something broke and then the eighteen hundred–pound boom fell," said Sheriff Hampton.

The worker's body was left suspended 240 feet in the air, and crews were left trying to determine how to get the man down. Crews with Northern Kentucky Technical Rescue finally removed the body around 10:00 P.M. Thursday.

"Our thoughts are with the family, loved ones, and colleagues of the worker who died in the July 2 incident near Cynthiana, Kentucky," said Verizon Wireless in a statement to WKYT about the incident. "We are working closely with our vendor and the sheriff's department as they continue to investigate the situation."

First responders struggling with the intensity of the accident scene have help available. The city of Cynthiana offers employee assistance and counseling for those that need it.

Fire Chief Jay Sanders says his firefighters talked after their shift to discuss any issues. They talked again twenty-four hours later.

"We don't get a lot of training on how to deal with horrific accidents, the stuff you shouldn't see in a lifetime, but we do, and that's why we stand watch," said Sanders.[51]

51. WKYT-TV, Lexington, Kentucky, "Cell Tower Death Identified as 28-year-old Father of Four," July 2, 2014. www.wkyt.com/home/headlines/Cell-tower-death-identified-as-28-year-old-father-of-four-265705051.html.

CHAPTER ELEVEN MISSISSIPPI
KIRK TO ENTERPRISE: CAN YOU HEAR ME NOW?

When I pulled onto Abbey Lane, I could not find a parking space. I had made the bumpy push-and-pull run from the Jersey suburbs to the Nassau County suburbs, leaving the land of Lou Costello, Bruce Springsteen, and jughandles for the land of Billy Joel and Zippy the Pinhead. Those were two of Levittown's claims to fame. Ellie ("It's My Party and I'll Cry If I Want To") Greenwich lived here for a while as did, some say, Bill O'Reilly—but I don't believe that for a second.[52] I expected some solace in familiarity there on Abbey Lane. Normalcy, perhaps, after a very abnormal morning. But the street I grew up on with its fenceless yards and hockey goals and a thousand Norway maples and zero cars was absent. Nassau County was fast becoming Queens. Suffolk County was becoming Nassau County. It was the natural ebb and flow of the city. Migration. Expansion. The only thing keeping The Hamptons from assimilation by the huddled masses was an entrenched army of money and zoning laws bristling like bayonets on the western bank of the Shinnecock Canal. The Long Island suburbs, once the airy cool retreat for the sweltering boroughs, had been used to their limit. That quota had been filled. If this island was

---

52. O'Reilly has claimed to live in Levittown, but this has been disputed.

never discovered by the Europeans, I have no doubt that today the Iroquois would be tripping all over themselves and whacking each other with lacrosse sticks while vying for space for their longhouses. There was no place to expand thirty years ago, and there was no place now. Five pounds of shit in a one-pound bag. The only place to go was up, and the new Levittowners were unrivalled in their building of "extensions," literally houses built upon houses creeping skyward off their sixty-by-one-hundred-foot lots, sided in colors challenging the spectrum, sequestered behind wobbly white PVC fences. Mini-McMansions all. As I crawled down Abbey Lane, I realized Levittown had indeed morphed from the sublime to the ridiculous.

As for historical value, Levittown was nothing less than the genesis of the America we know today. The only person who changed the physical, social, and financial face of America more than William Levitt was perhaps Henry Ford. Levitt, son of Abraham, was the first to apply Ford's assembly-line dynamic to the building of affordable homes on a massive scale. After World War II, recognizing the endemic need for housing for roughly eleven million returning veterans and their burgeoning families, William Levitt and his brother Alfred turned four thousand acres of recently blighted potato field into the nation's first true suburb. Between 1946 and 1948, they had erected 17,400 new homes. Misters Levitt planted nails. Popcorn houses snapped up straight. The city broke its wind, and Brooklyn sagged. Soldiers took their wives, kicked the wet dirt off the walk, stenciled numbers on the curb, and bought lawnmowers. Even at the rate of building up to thirty homes a day, the "rabbit hutches" (as the soldiers liked to call them) were sold before they were even completed. My father was part of the second-generation city exodus. Though I was born in Brooklyn twenty-two miles to the west, and I still have vague memories of that place (of thin hopeful trees rising from concrete, of Teddy's Italian Deli with fat salamis hanging from the ceiling and sawdust on the floor, of whole-headed fish lying in crates of ice out on the sidewalk, of the half-sunken vestige of trolley tracks turning the corner of DeGraw and Fifth), I was indeed *from* Levittown. And where I was from was no longer there. It was gone as surely as if a protracted tornado had swept through and scrubbed away the pasts of 17,400 families in a few short years. I cursed this because the new Levittowners, the invaders, would not allow me to park my frigging car in front of the house I grew up in.

When I was growing up, nobody worked *in* Levittown. Every morning most of the soldiers took the one family car to the Long Island Railroad Station up in Hicksville or down in Wantagh and rode the rails back to where they were born to make a living. The women walked. And the Levitts made provision for this. They positioned village greens with small shopping centers, Olympic-sized pools, parks, bowling alleys, and schools within walking distance of every home they built. The streets from dawn till dusk were as wide and empty to us as our imaginations would allow. And those streets were cool and dark with shade, Levitt having planted a minimum of six trees and copses of shrubbery on every lot. But between verticillium wilt and having been planted too close to the sidewalks, most of the Norway maples were dead and gone, most noticeably *my* Norway maple.

My Norway maple stood in the northeast corner of Dad's lot, and if you climbed to the very top and looked west, you could see the NYC skyline twenty-five miles away. You could see the World Trade Center being erected day by day, floor by floor, and that is where my dad went to work. I imagined I could see him, muscles straining in his jeans and white T-shirt, hanging the high steel, a no-filter Camel dangling from his lips. But nobody believed me when I told them this. I was ostracized for years in show-and-tell because nobody would believe me and my tree. Finally, I took up a Kodak 110 instamatic camera, and before my mom could yell *for the last time get the hell out of that tree*, I snapped a few shots and dropped them off at the Fotomat and then brought the prints to school and even Mrs. Lloyd had to admit I was right. From then on they believed most everything I said, which was both a good and a bad thing. Telling them I brought a dolphin back from vacation in Florida that was currently living in our swimming pool and eating Little Friskies was a true test of their faith.

I parked in a shopping center on Gardiners Avenue and walked to my brother's house, just a few hundred feet, still reeling from change. Nothing was as it should have been. Sammy's Inferno, the pizzeria where Sammy Esposito would actually sing opera (or sang what he presumed we kids would *believe* was opera) as he spun pizza dough in the air, was replaced by a Vietnamese nail salon. The A&P, which used to have coffee-grinding machines at the end of every checkout lane, was now broken up into a sushi joint, a liquor store, and a stationery store. The only recognizable feature was the wall of Evan's Army Navy Store, against which we would play handball until our knuckles bled. The post office remained there, however. Even invaders get mail.

Goddamnit, who did this? It was fucking depressing. Who took paradise and put up a parking lot? Some people will say Misters William and Alfred Levitt, sons of Abraham, did just that in 1946. Some say they created the Plasticine cookie-cutter America that I found as endearing as the Paramus malls. That they not only invented but perfected affordable mediocrity. I say bullshit. The Levitts didn't foul Long Island any more than the gas stacks fouled Bayonne. The City did that. New York City did that with all its unstoppable drive and unquenchable need. The City *gets* what The City *needs* and what The City *wants*. And each day it needed about 1.6 million (some estimates are as high as four million) commuters to fuel that want. Seven out of ten people pounding the pavement in NYC take the bus or the train or the ferries or the tunnels or the bridges to get there every single day. To put that in perspective: imagine that every morning, the entire population of Philadelphia or Phoenix got up, brushed their teeth, and went to work in New York City. And that every mother's one of them is carrying at least one cell phone.

The number of mobile devices rose 9 percent in the first six months of 2011, to 327.6 million—more than the 315 million people living in the U.S.

—*Washington Post,* October 11, 2001

It could not be more American that that America has more cell phones than Americans.

It is fashionable in the industry to say that the emergence of cellular technology is comparable to the Industrial Age, westward expansion, the transcontinental railroad, the oil boom of the early 1900s, or even the world-shaking advancements of Graham Bell, Tesla, Edison, and Ford with all the resplendent altruism and heroism that entails. But that would be both a gross understatement and demonstrably absurd. The *understatement* is that nothing has had a more direct, more all-encompassing, more moment-by-moment, second-by-second effect on Americans than the cell phone and its ever-changing and ever-adaptable technology. There *is no* comparison. Nothing even comes close. That is a fact I defy anybody anywhere to dispute. Personally, I hate the damn things because they are chronically unreliable and, in terms of quality, have yet to achieve the clarity and crispness of the landlines we were so dependent upon twenty years ago—those things we used to hang on the kitchen wall. Like crappy beer, this isn't about quality. This is about volume. But 315 million Americans can't be wrong, can they? And every Christmas many of them will wait in very long lines to get the next and newest cellular doodad with this G and that G, the one they just *know* will make all things in their life hunky dory, not knowing there is a damn good chance the promised advancements in speed and coverage most likely have not yet even been *installed* on many, many, *many* towers. But that snake oil notwithstanding, there is no denying the importance and permanence of the device.

What is *absurd* is that there is anything altruistic or heroic about what the industry does, what tower dogs do, or why we do it. We do it for the same reasons the carriers do it. To make as much money as we can. Period. We are not here to make anybody's life better. We are here to sell phones and provide the coverage that sells phones. Where it is true that when applied to emergency services, the cell phone has made a marked improvement to the arrival time of first responders, I often wonder what else it has truly improved, and at what price. Yes, the ambulance is getting there faster with the Jaws of Life to extract a mangled driver from the wreck, but the ambulance is also making a lot more of those runs because

> . . . *driving a vehicle while texting is six times more dangerous than driving while intoxicated, according to the National Highway Traffic Safety Administration (NHTSA).*[53] *The federal agency reports that sending or receiving a text takes a driver's eyes from the road for an average of 4.6 seconds, the equivalent—when traveling at fifty-five miles per hour—of driving the length of an entire football field while blindfolded.*[54]

So after they pry the driver from the mangled wreck, they can pry the cell phone from his mangled hand. Such is progress.

If we cellular providers and installers are indeed the dominant technological force of our times, that force can be broken down into the three basic components of all our technical revolutions: robber barons, consumers, and coolies. You are a consumer. I am a coolie. Our robber barons are faceless. They are Verizon, AT&T, Sprint Nextel, T-Mobile, Clearwire, MetroPCS, U.S. Cellular, Cellular South, ATN, and nTelos, to name a few. Even the Children of Andrew might be able to tell you what Ford and Edison invented, but ask anybody, including the eighty-five hundred to ten thousand

---

53. Todd Wilms, "It Is Time for a 'Parental Control, No Texting While Driving' Phone," *Forbes Business*, September 18, 2012.
54. National Highway Traffic Safety Administration, "Blueprint for Ending Distracted Driving." Washington, DC: U.S. Department of Transportation, 2012.

tower dogs working in the field or the hundreds of thousands of men and women comprising the entire industry from soup to nuts, *Who invented the cell phone?* and I'll bet my two-dollar bill they won't get it right. When I asked the Three Wide Men, it was Jimmy Tanner who came the closest when he said, "I think it was Motorola."

> *Motorola was the first company to produce a handheld mobile phone. On April 3, 1973, Martin Cooper, a Motorola engineer and executive, made the first mobile telephone call from handheld subscriber equipment in front of reporters, placing a call to Dr. Joel S. Engel of Bell Labs. The prototype handheld phone used by Dr. Martin Cooper weighed 1.1 kilogram and measured 23 centimeters long, 13 centimeters deep, and 4.45 centimeters wide. The prototype offered a talk time of just thirty minutes and took ten hours to recharge. Cooper has stated his vision for the handheld device was inspired by Captain James T. Kirk using his Communicator on the television show* Star Trek.[55]

"That was not a fantasy to us," said Dr. Cooper. "That was an objective."

He went on to say that "I have been a science fiction addict since that genre existed, back to Jules Verne. And before that I was into fantasy and mythology." To Cooper, Dick Tracy's two-way wrist radio was as much an inspiration as *Star Trek* ever was.

But Jimmy Tanner did *not say* Martin Cooper. He said *Motorola*. Ford wanted every American to be able to afford a car. Edison wanted to enlighten the world. There are statues of Ford and Edison and Graham Bell and Tesla and schools and boulevards and museums and universities bearing the names of our great innovators. Even the notorious robber barons of the Gilded Age, the Carnegies,

---

55. *How William Shatner Changed the World*, DVD, directed by Julian Jones (Boulder Creek, CA: Handel Productions, Mentorn, 2005).

Astors, Vanderbilts, Rockefellers, and Morgans, who amassed billions off the backs of unregulated immigrant labor, who built the great, shameless mansions on Long Island's North Shore just seven or eight miles from Levittown, are ensconced in the American pantheon of progress.

But there are no monuments to Motorola.

We don't feed the pigeons under the bronzened likeness of Dr. Martin Cooper. The technology and its creators are as unappreciated and as much taken for granted as the ether that carries it. But that don't mean it ain't profitable. And that don't mean there are not great, shameless mansions being constructed today at the expense of cheap itinerant unregulated labor and invaluable loss of life. The modern-day barons are unheralded, yes, but by design. The CEO, board of directors, and shareholders of Verizon and AT&T do not wish to be associated with the deaths of Maureen and Ted McWilliams's only son and Jerry Case and Kevin Keeling. They do not wish to be associated with David Huynh, who died from injuries he sustained falling from a man-basket attached to a boomer just south of Eugene, Oregon, on August 9, 2013. The carriers have placed beneath them layers of corporate insulation that defy any attempt at making them culpable for their contribution in creating and propagating the one, the only *deadliest job in America*. There are layers upon byzantine layers of general contractors and contractors and subcontractors and middlemen clambering over middlemen who *have* middlemen, and every one of them slicing away at their piece of the cheese until little is left for the people who actually perform the work, until Brody Kilfoyle has to tell me, "Until they start paying me more, I can't pay you more."

> *Carriers work through contractors and subcontractors rather than employing climbers directly, and carriers don't have employees on-site when accidents happen, making it difficult for OSHA to establish responsibility up the contracting chain.*
> —*Wall Street Journal,* August 2013

These layers form a magical staircase in which the shit only falls downward. The pressure to complete falls downward. The onus for safety and the paper trail for the same falls downward. The only thing that ever goes back up those stairs is money. Try to nudge a little bit of corporate responsibility up those stairs, and they turn into a litigious sheet of ice. PBS *Frontline* could not crack those layers of insulation, and NBC *Dateline* chose not to even try.

Dr. Martin Cooper is a fucking genius and by all accounts one helluva guy. If Alexander Graham Bell shrunk the world, then Dr. Martin Cooper atomized it. And I doubt he is concerned about monuments. There are no monuments to William Levitt either. His legacy, once hailed as a veritable wonder and the envy of at-the-time urban planners throughout the world, is now the butt of jokes and disdain or scholarly dissertations on how it all went so drastically wrong. What by as early as the '70s the Nassau County cops started to call the *white ghetto* has visibly become a white and black and Hispanic and Hindu and Islamic and Asian ghetto with half-dead lawns and five cars packed into spaces meant for no more than two. Thirteen Mexicans had moved into the house next door to my brother's.

"They are very quiet," Rob said. "Except on Sundays."

As I walked down Abbey Lane, I could not imagine thirteen people living in that Levitt house, but then I remembered the Czachors, the Polish Catholic family that lived behind us on Sparrow Lane, and they had at least thirteen in that house. So maybe things hadn't changed as much as I thought they did. And to my amazement there was one monument left to Abbey Lane—and by its very survival maybe to William Levitt. That lamppost with the busted access panel stood where it always had. Many of them had been replaced, but not this one. And much of the graffiti remained, complete with hearts and arrows and declarations of never-ending love, unrequited love, and wistful promiscuity. *Tom Gorecki Loves Laurel Hays. DD & DJ. Teddy* [indecipherable] *Holly. Shelly Tagliata puts out.* I had heard she did, God bless her, but I never had the privilege.

My brother Rob was not home. He had to head into the city. He left a note for me, a key to the back door, and $260 on the kitchen table. His wife was at work, his four kids out and about. I was so out of touch with my family I had no idea what they could be doing and wondered who owned the four cars parked bumper to bumper hugging the curb in front of their house. I sat in the backyard, and it felt cramped, though it never had before. All the blinds were down in the house next door, but I could barely hear the faintest Latino rhythms bumping against the glass. I stared at the patch of ground where my tree used to be and imagined myself up there, swaying atop the branches, peering to the west, looking for my father, inching his way up the freshly riveted steel of the World Trade Center.

Our familial connection to the Twin Towers didn't end with my dad. My sister worked there, and my brother Rob was in charge of all the elevators in the Trade Center complex, as well as most of Lower Manhattan, for Schindler. When I worked for John Grace, I was responsible for getting all the tools to the steamfitters who worked there during the HVAC renovations of the mid- to late-1980s and was there at least once a week. But I had nothing to do with the place THAT DAY. My sister was not there that day. My brother was in midtown that day. A few weeks after the towers fell, I visited my brother in a condo in Battery Park City where Schindler had put him up so he and his men could work on some small part of the untangling. We would sit on the tiny, two-person balcony at dusk drinking beer and watching the cutting torches shoot blue flame and white sparks all night long.

"What's the hardest part of all this?" I asked him.

"Boots," he said.

The site was still so hot, the composite soles of his crewmen's boots were melting on the steel and had to be replaced two or three times a day. "Can't get enough boots down here," he said. But we did not attach ourselves to that tragedy. We attached ourselves to the building of the towers, the operation of the towers, and not their demise. For too many people that was a real catastrophe, and it would be wrong to usurp that kind of loss with half-assed tales of obscure connections to *that day.*

My family found it hard to even use the term *9/11.* It was too simple. Too pat. Like *24/7* or *raise awareness.* You hear that shit enough, and it has no meaning. To paraphrase the English teacher Troy Boucher, who was probably paraphrasing someone else, *the first person to say "my love is like a red rose" was a genius, and the second was an asshole.* But thinking about the Trade Center did get me thinking about my dad and what he was and what he did and what he might have left me besides a knife and a two-dollar bill. Was I built for the high steel? Did I inherit some fucked-up gene that attracted me to that kind of danger? My father had over three hundred jumps in his jump book. He worked the steel on the tallest buildings in the world. He didn't discuss these things. He just did them. *Nobody ever woke up in the morning and said, Today I am going to climb cell phone towers.* Was I born to this? And if so, why the hell did I go to grad school?

I do not expect the carriers to be saints any more than they should expect the dogs to be saints. Henry Ford was a rabid anti-Semite. William Levitt would not allow realtors to sell homes to blacks or browns or yellows. Vanderbilt is credited with killing thousands of coolies during the construction of the Union and Central Pacific Railroads, but most historians will put the number at about 130. Still, that is quite an accomplishment because considering about twelve thousand workers built that railroad, one in every 92.3 of them died doing it, making it hands down the deadliest job of their day. Sixty men out of ten thousand workers died building the World Trade Center, averaging one in 166.6 tradesmen. Tower worker fatalities since January of 2003 average one dead dog per sixty-five workers.

I do not expect the carriers to be saints, but I do expect them to peel back a few layers of insulation when a worker dies, rather than what they do—which is to immediately attempt to find fault in the climber or the subcontractor or the contractor or the general contractor and cover their asses all the way home. Then they might order a *stand down* or issue a boilerplate press release expounding how the carrier has always a been a staunch proponent of safety in the industry, *whee, whee, whee, all the way home.*

> *Worker safety has always been a hallmark of AT&T . . .*
> *Though AT&T does not handle wireless tower construction*
> *itself, we strongly support the work of OSHA and the*
> *National Association of Tower Erectors,*[56] *who together*
> *launched their wireless tower worker safety initiative in 2007,*
> *resulting in a dramatic improvement in worker safety.*[57]

The stand downs usually last twenty-four hours and consist of your supervisor (Sarge or Jimmy or Scotty) rounding up the crews and saying please don't do what these assholes did and make sure you fill out your JSAs. It's Nancy Reagan saying *Just say no*. It is always too little too late, but it looks good to whatever press deigns to cover the incident. The *dramatic improvement in worker safety* perhaps is not that all dramatic considering that in the four and a half years before the initiative, there were fifty-seven fatalities, and in the nine and a half years after the initiative, there were seventy-three. That is an improvement, yes, but nothing for any industry to be even remotely smug about.

I am not raging against the machine. Raging against the machine is not only fruitless but also paranoiac. Thinking the machine killed these tower dogs is just plain foolish. But it would also be foolish to think the machine does not exist. And it would be irrational to think that the machine does not directly contribute to these deaths. Because it does. It does because, despite any profession of *safety first*, the primary general order of all tower work is *get it done now*

---

56. Or NATE. And we will get to them later.
57. Statement issued by AT&T to PBS *Frontline*. Attributed to Mark Siegel, AT&T spokesman.

because for every second that system is not online and transmitting and receiving data, the carrier is losing money.

Jon McWilliams's parents believe *There needs to be some new regulations or laws. Maybe just basic companies caring. There needs to be some accountability*, and I believe that, too. Though there is no doubt in my mind the carriers really could not a give a shit about the baseline worker or his death, the fault lies indeed not (to maul Shakespeare) in their stars but in our own. But the reason it lies within our own selves I *do* place at the feet of the carriers because in my opinion, it is they who control the single most contributing factor to deaths in our industry—and that is that you *get what you pay for*, and carriers spend most of their money on all the wrong things.

Carriers spend the lion's share of the financing dedicated to any cellular build-out or upgrade *not* on the companies and the crews who perform the work, but on the multilayered wedding cake of middle-management contractors, consultants, and third-party auditors that may pocket as much as 50 percent (some say more) of the original cost of the endeavor before a single antenna is hung. The business model for this is sound, because utilizing those management firms enables the carriers to make drastic decreases to their own in-house workforce, both cutting their bottom line and passing the millions of dollars of cash layout for any given site on to the contractors, not to mention the liability. Shit, *especially* the liability.

Still, K.M.C.A. does not need that management, nor do hundreds of companies like them. Give them the blueprints and the material and the cut window and get the hell out of the way, and it will get done cheaper and faster than when forcing installation entities to navigate the policies, protocol, and the absurd and redundant reams of paper heaped upon them by those levels of middle management.

Never before in the annals of American industry have so few done so little and for so much money as the middle contractors that separate the carriers of cellular service from the task and the crews in the field.

More money for the installation contractors who perform the work means more money for their workforce. And one might wonder, *Even if they got more money per project, why would they pass it on to their workers?* Because they want to, and they have to. This industry needs the kind of worker it already professes to have but largely does not. K.M.C.A. and companies like them spend thousands upon thousands of dollars on equipping and training their men, and it can take months and even years for a worker to become a competent and profitable member of the team. But few make it that far because inevitably, they realize they can indeed make as much money per hour behind the register at a QuikTrip, and also, the skills they have attained are far more marketable to—guess who?—the middle-management firms that suppressed their wages in the first place.

AUGUST 10, 2014

Thomas Lucas of Toledo fell eighty to ninety feet while painting a communications tower south of Stockton at 13072 E. Moresville Rd. OSHA investigated the fatality and cited Sherwood Tower Service for one serious and two willful safety violations. The agency has placed the company, which specializes in communication tower painting and maintenance, in its Severe Violator Enforcement Program. "Three children are without a father because of a preventable tragedy," said Jacob Scott, OSHA's area director in North Aurora. "No one should have to endure such a painful loss ever. Inspecting and making sure protective equipment is in use and working properly is a common-sense safety procedure that saves lives and prevents injuries. Companies that ask their employees to work above the ground have a responsibility to provide adequate fall protection to workers. OSHA has seen a disturbing trend in preventable deaths and injuries in the telecommunications industry." [58]

58. *Wireless Estimator*, "OSHA Cites Wireless Contractor with $114,800 Fine Following Painter's Death," August 10, 2014. wirelessestimator.com/articles/2015/osha-cites-wireless-contractor-with-114800-fine-following-painters-death/.

CHAPTER TWELVE MISSISSIPPI
I HAVE CHARTED A COURSE TO THE VINEYARD

Leaving Levittown, all I could think was *Up yours, Thomas Wolfe, you prescient bastard*. I was thinking about the good old days that weren't so good. Crossing back over the George Washington Bridge and into New Jersey, I remembered that first tower in El Dorado, and when I remember El Dorado as somehow being the good old days, I never feel the cold. The cold. The heat. A broken heart. The loss of a loved one. You can't feel those things like you can when you are mired in that time and place any more than you can really feel anyone else's pain. You can *say* you feel it, but you can't *feel* it. You can't feel it because it ain't yours and it ain't now. But what I did feel was a melancholy sense of loss. The loss of working with men like Power and Hangman and Brody. The work will always be just that, but one of the things that makes it doable is the men you are with. You spend more time with these guys than with your family, your woman, or your children. They are the first thing you see in the morning and the last thing you see at night, and though it is often said *you don't have to like him but you do have to work with him*, that seldom floats. It is impossible not to take the job home when the job *is* home.

When I rolled into the parking lot at the Extended Stay, my thoughts had been so filled with what I wanted to be the good old days that I could not remember the drive over. I had arrived on automatic. And for the most part, that is how I went to work for the next few days. There is a reason men in the field have the option to take off for a week every six weeks, and I was the poster dog for that at the moment. I had been out over nine weeks. I was getting sloppy. I was getting tired. The rush of being back among the men had waned. The nightly debriefings were turning into postmortems. My intolerance of the new hands was thickening by the day. I was getting pissy. Jimmy and Scotty had mobed back to Nashville to work the market from that end, and Sarge and Devil Anse, already burning out, had much more on their plates. And most of all I missed Meagan and the baby more than I ever thought I could. Phone calls weren't doing it. Phone calls were not enough. I had almost convinced myself that it was okay. That my son was just an infant and he could not possibly remember that I was not there. He could not hold that against me. Before he was born, I had told my sister that I was afraid I did not have the "Dad Gene," and up until the moment he arrived, I was still unsure and petrified. But the moment he was born I had that life-changing bullet between the eyes I had been warned about by those bleary-eyed sages who did have kids, and damned if they weren't right. I wanted to hold my kid. I wanted to sit in the rocker with his breath on my chest and the back of his head in the palm of my calloused hand. Hell, if I left right now, I could be there the following morning, and that thought briefly skittered across my mind. It didn't hold, of course. It was flattened by the fact I had to feed the little thing. I just wanted to be there while he ate.

I found Super Mario, thanked him, and gave him his money. I checked in with Daryn for my marching orders, and he said we wouldn't know shit until muster. But I did find out that SeanDog had been permanently placed on our crew, as per Sarge, so he *could learn.* This did not bode well. In the next few weeks he was in market, SeanDog was kicked out of every bar he stumbled into for fighting or for being a general nuisance. He loved to break glasses. Heavy mixed-drink tumblers would somehow manage to crash to the floor whenever he was around, even in the parking lot. And he loved to "play-wrestle" with the guys, sneaking up behind them and slapping them in a hold, leaving them no choice but to tumble about the parking lot with him. And, to complicate matters, he was one big strong sonofabitch and knew it.

I went back outside and sat in a folding chair and watched the mosquitoes and dragonflies dive-bomb Lake Scum. I thought of Laurel and how I'd have to tell her about the lamppost. I could hear her voice saying *GORECKI?! OH MY GOD!* I was the only one in the lot. And my left turns and synapses began to swirl inside my head like an unbalanced dryer load with heavy wet sneakers knocking the door wide open so you'd have to get up and reset the machine over and over again. It was all like that here in New Jersey. It was all out of sync. *I* was all out of sync. And though I can't say I KNEW, I can say I had a very strong feeling that all the bozos on this bus were just accidents waiting to happen and I did not want to be around when they did. I needed to take my break and recharge. I needed to take my break and hope that when I returned, I would be assigned to a different market.

The Greeks had a word for us. All of us. If one judges action A to be the best course of action, why would one do anything other than A? Why would Frogger play Frogger? Why wouldn't they get rid of SeanDog? Why would I free-climb twenty feet of steel in the Amboys? Why would all of the eight men who died free-climbing so far that year free-climb? Why would anybody do this fucking job? And after my break, why would I return to it?

Any juice I had left was drained away by having to work with the two new dogs on my already ad hoc crew, SeanDog and Big Ben. Whereas SeanDog was a never-ending cause of worry, Big Ben was just plain worrisome. We were sent to Rahway. A rooftop. Normally a rooftop installation was a welcome change from towers because you could actually stand on your feet the way nature intended you to. Sometimes you'd have to hang off the side of a building, *window-washing*, we called it, but that was okay too because gravity worked in your favor. You would lower yourself on ropes to your workstation (usually antenna mounts attached to the side of the building, or if they weren't there we would attach them), and when you were done, you could rappel[59] to the ground and take the elevator or stairs back up again. But with SeanDog and Big Ben along, the change was anything but welcome. Big Ben was not at all a bad kid. He was big and blond with a Tom Joad haircut he had kept from his days in the army. He was a Gulf War vet, a recipient of several Purple Hearts with the scars to prove it, scars he tried not to draw attention to. If Big Ben wanted to talk about his war, that was his prerogative, and he did not. Any damn fool who read *the primer* would know enough not to pry. So when on our first ride together to the twenty-two-story apartment house in Rahway, SeanDog bombarded him with inappropriate questions:

Damn, homes, you get them scars in Iraq? Fuck they tore you up good.

You ever kill anybody? You ever blow them towel-head muthafukas away?

Big Ben pretended not to hear him. He didn't need that crap. He looked at me with his broad ruddy face like a child looking to hide behind his daddy's leg. I decided right then I would sponsor him without his knowledge, help him negotiate this new minefield he was tiptoeing through. I tossed Big Ben a look that said *neveryoumind that asshole*, and to my surprise Big Ben understood instantly.

---

59. *RAPPELLING* is no longer allowed. We practice what is called a controlled descent.

That cheered me up a bit. Like Hangman and Jess Pulaski, this was a man of few words, and dammit if we didn't need more of that around the Extended Stay.

From the rooftop of the apartment house, all that Rahway had to offer lay beneath us, a panoramic microcosm of many Eastern Jersey towns. At the epicenter was us and the urban decay of the old downtown, red-bricked and sooty, but spreading out from the middle the neighborhoods softened as they grew more prosperous. We could see tree-lined avenues with small apartment buildings and then two-family duplexes and then single-family homes. The Rahway River, shallow as Lake Scum in places and punctuated with old tires and mattress springs and shopping carts, pressed dark and dirty out to the Kill Van Kull and the Elizabeth stacks and Staten Island. Directly to our south on Rte. 514, we could see Rahway State Prison, a hulking behemoth of penal deterrence. Built in 1896, its design was incongruous. The focal point was a huge yet squat tiled dome resembling a mosque, and from there jutted a pair of two-story redbrick barracks at forty-five-degree angles. To everyone in the Northeast, when you said *Rahway*, you said *prison*. When you said *Rahway*, you said *house of horrors*. The association was so entrenched that one hundred years after it was built, the residents of Rahway lobbied the state for a referendum and had the name of the facility officially changed to East Jersey State Prison.

Daryn, Big Ben, SeanDog, and I walked the roof. Vic with a D came along to the site, too, but he was down in the trailer doing whatever, and I had long given up on him being any help anyway. Daryn and I improvised a plan. I would "run the top" and he would "run the bottom." Somewhere in the bowels of the basement of this apartment building was a "site house," and he would have to find it, and between the two of us we would figure out how to best go about this job. The site was not playing, and my first task was to decom over one thousand feet of old 1⅝–inch coax (lying flat on the roof in cable trays) and get it off the roof. There was no way we were going to get it through the roof hatch and into the elevator. So while I had SeanDog and Big Ben cut up the coax into manageable eight-foot lengths, I erected a boom and a block and fifteen hundred feet of rope at one end of the roof, from which I would simply lower the bundles of cable to the ground. It was a little tricky because directly below the boom was an exit from the first floor that led to a small playground and an empty swimming pool, and the tenants were constantly coming in and out. Because of this I decided it should be a three-man job. I would hook a tag line to the load. While SeanDog hooked up the hundred-pound bundles on the roof, Big Ben and I would work the ground, me on the load line and he on the tag. Simple enough.

But we still had a few hours of cutting and wrapping to do. Every time I turned around, SeanDog was nowhere to be found. If there was a place to hide, he would find it. In the stairwell. On the loading dock. In the parking lot. In the trailer. In the truck. He was never where he was supposed to be, doing what he was supposed to do, as if my very precise and direct instructions to him were only a suggestion. And always, *always* with his cell phone plastered to his ear. It got so bad I took a page from Sarge's handbook and, following the chain of command, I gave Daryn my cell phone and said, *Look, take everyone's cell phones and put them in the truck. There is nobody we need to be talking to. If it's work-related, they'll call you. Then maybe we can get some work done.* Daryn said, *I'll talk to him.* But it didn't do any good. Even when I could nail him to a task and stand over him like a suspicious proctor, SeanDog worked one-handed, his other hand clamped around that phone. This wasn't some teenager. This was a grown man who literally spent his whole workday texting and talking. And though he was the worst I had ever seen at this, he was not the only one. Keeping guys off their phones had become a very real problem. Memos were flying out of Navaho Road:

Any employees texting or talking on their cell phones while operating the cathead will be fired.

Any employees texting or talking on their cell phones while driving a company vehicle will be fired.

Any employees texting or talking while ON THE TOWER will be fired.

They might heed the directive for a few hours, but once they realized nobody was getting fired, they backslid right into the same old shit. At the least, it slowed down the work. And of course it was dangerous as hell. Somewhere, somehow these guys got it in their heads that the mobility of their phone made it okay to use it all day long. This was not a convenience for them; it was a constitutional right. While we were doing a heavy stack in Mississippi, Sarge got so fed up with watching guys talking on their phones while we were flying steel he took all our phones, tossed them in his truck, and responded to the protests by saying, *If you die, I'll call your mother.* Some guys had company phones, and the abuse could be even worse because they knew they were not paying the bill. One jamoke in Atlanta ran up $470 in charges on the Psychic Hotline. Brody kicked in his motel room door, grabbed him by the shirt, and said, "I'll tell you what your future is, waterhead. Your FUTURE is you're gonna pay this fucking phone bill."

If it were my crew, I would have sent SeanDog to the truck without pay or per diem and let him talk all day. I would let him sit in the motel without pay or per diem and let him talk all day. And when he got his phone bill and realized he could not pay it, maybe then he would realize this was a job. Maybe that was the solution, I thought. Maybe if I were running a crew instead of settling for one, I would not be so miserable. There were things I had to learn about this market (the new plumbing diagrams were a nightmare), but I had led crews before. Why sweat the morons like SeanDog? If I was crew leader, I could fire his ass. I would talk to the Three Wide Men about that that night. Right then, I had more pressing problems.

I was beginning to think that Big Ben might be damn close to deaf. I would call him over for this or that, and it took three or four times, my volume rising with each beckoning, for him to acknowledge me. If he did have a hearing problem, I would have to know that because when we got him up a real tower, he would have to be able to communicate. A bigger concern was that several times that first morning, I noticed him just staring off into space. Daryn noticed it, too. "You think he's got that post-traumatic shit going on?" Daryn asked. He was less than fifty feet away, and Daryn and I took turns calling out his name, and not once did he even flinch. Daryn went up and put his hand on Ben's shoulder. He did not start. He did not freak out like Daryn was half expecting him to do. He just looked over at Daryn and smiled.

"Yeah?" he said.

"You okay?" said Daryn.

"Yeah," Big Ben said, and he knelt down and went back to work on the cable tray.

Daryn took me by the elbow. "Keep him away from the edge of the roof," he said, his face loaded with worry. I would do that. Several more times Big Ben left the rooftop, went some other place without going anywhere. I could only speculate that place was filled with sand and blood. At one point I did what Daryn had done and placed my hand on his shoulder and quietly said, *Ben?* He turned to me and said, "Is that the Empire State Building?"

"Sure is," I said.

His awe was childlike, and I took a minute to point out the Chrysler Building and what I still called the PanAm Building, the major bridges we could see, and the general layout of the boroughs. I even pointed out the Williamsburg Clock Tower. "I was born a few blocks from there."

"Wow," he said, "the Empire State Building." He was the first of the Southerners up here to show even a small appreciation for the greatest city ever built, and that in itself made me like him even more. It reminded me of my first trip to L.A. when I got on the phone to Tony Bill and said, "Guess what? I can see the Hollywood sign!" as if I were the only one among three million Angelinos who had that vantage point.

We ate lunch down at the trailer and lined out the afternoon. Daryn and Vic with a D would work in the site house while me, Big Ben, and SeanDog would lower the twenty or so bundles of coax from the roof to the ground and then load as much as we could into the trailer. Daryn inspected my boom rig and approved. I went back up to the roof with SeanDog and Big Ben and walked them through it.

> To SeanDog: *You hook up the load like THIS. When I WHOOP, you hang it over the edge of the parapet. I've got the load. Just keep your eyes on it all the way down. Ben will tag it out. Don't worry, Ben, I'll show you when we get down there. Sean, after we unhook the load, we'll send the rope back up to you. KEEP YOUR HANDS ON IT. Don't let it just fly up.*

SeanDog said, *Hang on a sec, Babe,* into his phone then looked at me impatiently and said, *Yeah, yeah, I got it, I got it.*

On the way down to the playground and pool area, Big Ben and I joined Daryn for a moment inside while he showed us we would have to drill about twenty six-inch-diameter holes through twenty-two floors of ten-inch concrete without creating any dust for the tenants to complain about. While pondering that impossibility, Daryn's phone rang, and he stepped into the corner to answer it.

"You did WHAT?" he said, and he ran for the exit.

Big Ben and I followed. That was SeanDog on the phone, and what he had done in some senseless spate of initiative was throw thousands of pounds of carefully wrapped bundles of coaxial cable right off the rooftop and down twenty-two stories into the court-yard between the first-floor exit and the playground. When we got to that side of the building, we could see a thirty-foot-wide rat's nest of exploded bundles of coax. Having to rebundle the whole shebang was not even the issue. In the playground, a few wide-eyed kids sitting on their swings just stared at us as if to say, *You done now?* And Daryn, finally, went apeshit.

This was unacceptable, he explained to SeanDog when we got back on top. "THIS is not only the kind of shit that will get us kicked off this job; this is EXACTLY the kind of shit that will get this whole fucking company kicked out of market! Do you understand how serious this is! DO YOU?!"

"What's the big deal?" SeanDog said. "Only took a second."

"THE BIG FUCKING DEAL IS THAT YOU COULD HAVE KILLED SOMEBODY! What would you have done if someone walked out that door?! You can't STOP this shit once it goes! This isn't the fucking movies!"

It was not sinking in. SeanDog just looked at Daryn, completely unfazed. Not only did he not understand, he simply did not care to understand. He wanted to get back to his phone call. Daryn stared at him, mouth agape, challenging him to come up with some reasonable explanation. SeanDog looked back at Daryn and shrugged. "What's the big deal?" he said.

"That's it," said Daryn, stunned and frustrated into submission. "Wrap it up."

On the ride back we stopped and got some beer, and SeanDog, on his phone to Babe, was slamming them down and crashing the empty bottles to the pavement outside his window. I told him to knock it off. Vic told him to knock it off. Daryn told him to knock it off. SeanDog said, *What's the big deal?* That night in the parking lot, I saw Daryn and Sarge off in the corner by the welding area, no doubt discussing SeanDog's performance that day. Big Ben and I decided to share a room, and I watched out the window. The more Daryn's arms flailed, the more Sarge's arms tightened across his chest. I knew what Sarge was saying. He was telling Daryn that no matter what the fuck the guy did, it was Daryn's fault and Daryn's responsibility. That what SeanDog did was on Daryn's watch, and that was that. It did not matter that the man just refused to listen to anything he was told. It was Daryn's job as crew leader to get him to listen. To get him to straighten up and fly right. If you can't run your men, what the hell use are you? I decided against talking to the Three Wide Men that night about stepping up to crew leader. SeanDog wasn't getting the dog he so deserved, and that was enough for me to know that whatever responsibility and power a crew leader had was useless without the support of the bosses. Sarge would talk to SeanDog for sure. And SeanDog would say, *Yes boss*, and then do whatever the hell he wanted to. He would defy Sarge himself, and that was as concrete a sign of insanity as I needed to know. I knew we would have to get rid of him another way. And I devised a plan in my head. *If we can get SeanDog down to the Bing when Sarge is entertaining the guys from GeoDyne Tech and SeanDog just behaves as I know he will, then maybe . . . just maybe . . .*

But that never came to pass because SeanDog would take himself out of the mix as surely as Frogger did. Till then, my way of dealing with SeanDog was not to deal with him at all, and I was catching hell for it. If they sent him up the tower with me, I would send him down. If they complained I was taking too long, I would say I am doing the best I can since I am working alone. And I was. Humping antennas and mass pipes and coax, working all three sectors. This wasn't about dislike. SeanDog was unsafe on the tower at any speed, refusing to follow the most basic rules. I tried to utilize him on the ground, but I could not even trust him manually with a rope, let alone a cathead. Everything he rode up or down was too fast and too jerky, sending thousands of dollars of antennas crashing into the tower steel. His head was not there. He was in high school waiting for the bell to ring. Big Ben and Vic with a D had been parceled out to other crews, and some weird science prompted the bosses to make a crew out of me, SeanDog, and John Cooper, Bo's dad, of Beaumont, Texas. No crew leader to speak of, just a crew. And I decided after a few days on a tower with SeanDog I would do it alone rather than take the chance of this guy killing me. I would do as Pulaski or Hangman would do. They never caught hell for it, so why should I? But Sarge wanted to throttle me as much as White Chocolate. When he told me to take SeanDog up the tower anyway, I refused. I let him sit in the job trailer all day. John Cooper, recently brought to market by Jim, was a saving grace. Though his background was mostly civil, he was an experienced climber, competent and safe, excellent on rigging and the cat. He was one half of his son Bo, but wired like Devil Anse. His slight build and scraggly half beard belied his sheer strength and know-how. He was sixty-six years old and worked as hard as any of the young pups. And he was always on the radio asking, *You need me to come up there? I can, you know.* I did need him to come up, but I needed him more on the ground. The towers were getting done but way too slowly, and I knew we were probably losing money, but I did not see I had a choice. A serious accident would cost Brody Kilfoyle and K.M.C.A. a damn sight more than a few over-budget installations.

At the Extended Stay, things calmed down a bit. Though it was barely September, the Christmas push was coming on, and all those people standing in very long lines for their new gadgets would not be denied. We were working long into the dark, and that cut down on the intensity of the debriefings. Things were so uneventful in the parking lot we were almost becoming respectable. Guys would roll in, clean up, have a beer or two, and call it a night. Big Ben was a considerate roommate. He would grab a bite, shower, call his wife, and watch TV. Sometimes he would go to that other place where he got his scars, and I just let him go. The only reference he ever made to his war was when he finally answered SeanDog's query as to whether or not he had killed anybody.

"That's something we don't talk about," Big Ben said.

Whether he meant WE being he and his combat comrades or WE being he and SeanDog, I did not know. SeanDog sniffed, "Be that way," and finally stopped asking. But he did not stop being SeanDog, and the only real disruptions we had to contend with since Frogger always led directly to him. On one night, he engaged an exhausted Daryn in one of his sophomoric impromptu wrestling matches. Before Daryn could say, *Not now, dammit*, SeanDog had slammed him onto the pavement and cracked open his head. Cady saw this, dropped his skateboard, and hit SeanDog so hard in the jaw he knocked him out flat. All I managed to see was SeanDog splayed out in his stark white T-shirt and Cady helping Daryn to his feet. Little Cady. My hero. SeanDog's embarrassed excuses in the following days were that *Cady coldcocked me, I didn't see it coming*, and the general response from all was, *Neither did Daryn*. But even that did not enlighten SeanDog.

If you really wanted to condemn a fellow worker, you did not call him a muthafucker or an asshole or a slap-dick. There was only one phrase that seemed to make a dog stop in his tracks and reassess himself, and that was *That's not cool, man*. That's not cool. That's uncool. Dude, not cool. Uncool, dude. Say that with a disappointed shake of the head, and most dogs took notice and adjusted. Most dogs would wrangle up and say, *I'm sorry, you're right*. The most uncool thing SeanDog did among us was to Big Ben.

One evening when Big Ben was sound asleep, SeanDog burst into the room screaming like a banshee, jumping up and down on the bed and slinging lamps and phones and luggage. He then came sprinting out of the hotel and into the Olive Garden. When Ben came out into the lot, where I was cleaning out my rig, he was trembling. Some of the other dogs were roused by the commotion and inquired as to what had happened. I am not sure even Big Ben knew. But I could tell he was shaken and trying to hide it. His eyes were twitching in his head, and his fingers were shaky. His breathing was shallow and arrhythmic, as if he could not get enough air into the right places at the right time. If he had gone to his other place, he did not go there intentionally. It was SeanDog that sent him there, and I knew he *did* do that intentionally. It was all a big joke to him, and even after he showed up later and was told by every dog who saw him that was *very uncool*, it remained little more than that joke. He was just kidding. *What's the big deal?*

The big deal came a few nights later in a bar that did not know him. I was not there. None of us were. It took a few days to piece it all together. Our first inclination that SeanDog pulled off a big one this time was when the cops came to the Extended Stay. They were not the Ramsey beat cops or the state troopers. They were plainclothes detectives. And they were serious. They talked to a few dogs in the lot. And when they went inside to speak with the desk clerk, SeanDog came hurtling out his second-floor window and hit the ground running. That was the last time most of us ever saw him. The detectives were looking for SeanDog because he had almost killed a man. Jimmy Tanner would later tell me that SeanDog took one of his heavy glass tumblers and smashed it into the face of a bar patron. A man he had not spoken to and who had not spoken to him. "He was cut up really bad," Jimmy would tell me. "Was in the hospital a long time." Why SeanDog did this only SeanDog knows, and the detectives didn't care to know. SeanDog was not entitled to reasons. They wanted him bad, and though it would take weeks, they tracked him down all the way to South Central Florida, and he was arrested and extradited and convicted and sentenced to three hots and a cot, courtesy of the New Jersey State prison system. Despite the referendum from the lawmakers in Trenton, I hoped it was the penitentiary everybody still called Rahway.

Not too long after that, Big Ben quit. I was sorry to see him go. He was smarter than us in that rather than hang around for months getting paid for a job he did not really want to do, he made the cut clean and quick. That was respected by all. *No shame in it.* I thought about him on my drive out to Martha's Vineyard because it was he who inadvertently sent me there. It was his gawking wonder at the Empire State Building that got me turning left and reminded me of my own gawking wonder at seeing the Hollywood sign and of Tony Bill and Marc Shmuger and a dozen or so other contacts I still had back in the life I had abandoned (or that had abandoned me) to become a dedicated *for-real-this-time* tower dog. That was weeks and weeks ago, and I was already starting to think, once again, and despite my dismal record, that maybe I had one more project left in me, one more big paycheck. Maybe this time it would all work out. And something else was working here. After making a few calls, I discovered both Tony Bill and Marc Shmuger were within driving distance of the Extended Stay. Tony was at his summer home in Connecticut, and Marc was vacationing on Martha's Vineyard. Were the stars aligning for me? Did I dare hang my hat on another pitch? I had no pitch, but that had never stopped me before.

I first went to see Tony Bill. He was heading back to L.A. the next day, so we barely had time for lunch. He and his five-year-old daughter and I met at a rural tavern in northwestern Connecticut, a house of white slats, green gables, and thick mahogany table booths. A wagon-wheel joint across from a rustic train station. Every time I saw Tony, he was clad in the same soft and lightly worn white collared shirt, sleeves rolled up just below his elbows, and tan Docker-like pants. While his daughter carefully caromed the brightly colored balls atop the pool table, Tony leaned back, pulled his thick black hair out of his eyes, and said, "So how the heck are you?"

Even by SoCal standards, Tony Bill had a reputation for being the most laid-back man in America. He moved with a confident certainty, not slow but never rushed. He talked with a wry smile that said, *This too shall pass, and if it don't, the hell with it.* When we worked together, he always seemed to know when to turn me loose or when to rein me in. Sitting across from him, drinking iced teas and chewing on grilled fish sandwiches, I realized I had never even heard him raise his voice. I could see why this carriage was welcome on film sets, where too much drama and overinflated egos all too often held sway.

But that laid-back persona belied his endeavors as much as Devil Anse's size belied his strength. When it came to the film business, he reminded me of Jimmy *Godfather* Tanner. There was nothing he had not done and nobody he had not worked with, from Frank Sinatra to Brad Pitt. The game Six Degrees of Kevin Bacon could easily be played with Tony Bill instead. But I did not have stars in my eyes; I never did. What I had was respect because before I had ever met him, before he called me on my wall-hanging kitchen phone one day out of the blue and said, "Hi, this is Tony Bill," and I said, "Yeah, right, who is this really?" I had already been a big fan of his work. His staying power in a world that discards people as easily as medicated wipes was astonishing. And that was to be admired. He did not wear the biz on his sleeves any more than he flaunted his Oscar. He drove a Volvo station wagon. And in his glove box I bet he kept gloves. His Venice offices had more of the feel of a cottage boarding house than the always-next-trend décor of so many of the offices in which I used to peddle my wares.

We kicked around some old war stories. *Whatever happened to this project or that? Who died? No shit?! He still at Columbia?* I wasn't selling anything that day, just testing the water, and he knew that. He could tell when you had a project to pitch or were just floundering, and he knew I was floundering. He didn't offer me work. He didn't offer advice. He seldom did, and that was, in a way, appreciated. But being a good director, he was a good noodge; he could ease you into a direction without you knowing it, and when you got there, you thought, *How could I not have known that?* And this he did as he folded his napkin, scooped his daughter up in his arms, looked me dead in the eye, and said, *You really should be working on something.*

After they left, I sat in the parking lot at the train station for a while and watched a few of the commuter trains roll in and out, picking up and discharging no one. I opened my atlas to secure my route from there to Massachusetts. I put in a CD from the Billy Joel box set Power had given me for Christmas a few days after we left El Dorado. I knew I would see Tony Bill again. I knew I would see him soon because I had decided I was going to give it one more shot. And I knew I would arrive at his offices with just a few days' notice and he would make phone calls and I would go to meetings with people who would make more phone calls and I would not leave, not this time, until I was somehow, some way, someone else.

I would head southeast into the early New England dusk. I would listen to Billy Joel all the way and think of my father saying *Be who you become*, and in my head a skirmish was being waged over whether or not it was okay to say screw it to what you have become and try, try again to be what you want to be. Try until you are dead. *Don't give up that goddamned ghost, Delaniac.* Not yet, by cracky.

I would head southeast toward the Long Island Sound and then bear left and ride the dark Connecticut shoreline all night long, jockeying for space amid the endless procession of big diesel semis buffaloing their way up the I-95 corridor to make their morning deliveries in Boston and Bangor and beyond.

DECEMBER 19, 2014

Authorities have identified the tower technician who fell approximately one hundred feet from a cell tower in Greeneville, Tennessee, yesterday as 44-year-old Allen Lee Cotton of Sumter, South Carolina. Two of his coworkers from Central USA Wireless did not see him fall, according to Greeneville Police Officer Shawn Hinkle's report, which provided information regarding the 12:25 P.M. incident. "Mr. Cotton was wearing safety equipment, but it is unclear whether it was working properly," the report said.

Coworker Michael Gooch was also working on the opposite side of the tower and did not see him fall, according to the report. Another coworker on the three-man crew, Richard Wright, was working on a computer in his truck at the time of the incident, which remains under investigation by Greeneville police and the Tennessee Occupational Safety and Health Administration.

Wireless Estimator has identified that the workers were performing a project for Ericsson for their client, Sprint. The 197-foot monopole is owned by Crown Castle International.

Cotton is the tenth technician that was performing professional elevated services to die in 2014.[60]

---

60. *Wireless Estimator*, "South Carolina Technician, 44, Is 10th Industry Worker to Die This Year," December 19, 2014. wirelessestimator.com/articles/2014/12/.

It is not known why 28-year-old Stephanie Gurney of San Angelo, Texas, was on or fell from an SBA Communications five hundred-foot tower in a rural area ten miles east of Eden on Saturday at approximately 11:15 A.M., but it has been established that her boyfriend, Stephen Butler, who met Stephanie in September of 2014, was a tower technician for an Abilene tower contractor and started climbing in 2012.

The young woman died after falling from approximately forty feet, according to the Concho County Sheriff. Both the Sheriff's Office and OSHA are investigating the fatality.

A rash of rumors on Facebook and tower blogs that said that Gurney (whose occupation on her Facebook site said she was a housekeeper at a Super 8 motel) said she was being evaluated by Butler to see if she might be a viable candidate for climbing, or that she was working for a tower company, have not been substantiated by any authority at this time, according to SBA Vice President of Risk Management David Sams.

Butler did not respond to a request for additional information.[61]

---

61. *Wireless Estimator,* "Woman's Death Linked to Her Boyfriend's Tower Tech Profession in Texas," March 28, 2015. wirelessestimator.com/articles/2015/womans-death-linked-to-her-boyfriends-tower-tech-profession-in-texas/.

CHAPTER THIRTEEN MISSISSIPPI
A DECEPTIVELY SIMPLE TALENT

When the sneaking fingers of dawn crept through my windshield, I took it as a threat. I had arrived at State Pier in New Bedford, Massachusetts, just a couple of hours earlier and promptly fell asleep in the driver's seat. When the scarlet and purple chaos of the sky hit my eyes, I shot awake, thinking I was still at the Extended Stay motel in Ramsey, and I was relieved when I remembered I was not. I was dying for coffee, but being Sunday morning at 5:45 A.M., there was little hope of that. The ferry terminal was not open yet, so I shifted out of the Explorer and sat on the topped-off end of a pylon a few yards away. Before me sat one of the many inlets to Buzzards Bay and the enormous ferry dock. Behind me sat the part of New Bedford that cuddled the shoreline: dusty brick, two-hundred-year-old factory frontage, ancient duplexes coated in bright red or deep black. To the left and right moored trawlers and rusted hulks and tugs and coalers bobbed slowly in the subtle lapping of the tide. And all of this was shrouded in a fine, glistening mist. I closed my eyes and sucked in the salty spray and the skraking of the gulls who were just now awakening, too. It was both lonely and magical, and I sat there never feeling so alone and so invisible to the rest of the world.

When he was five years old, my little brother Brian discovered the secret to invisibility. He would close his eyes, and the world as he knew it would go away, and in his mind so did he. He did not boast of his newfound powers. He kept it to himself because the sole purpose of being invisible was to not be seen doing all the things he knew he should not be doing. He was invisible when he let my hamster out of the cage. He was invisible when he pulled my model airplanes off my chest of drawers and made them fly, something they were not meant to ever do. When you found a fractured Curtiss P-40 or P-38 Lightning at the base of the stairs, or you slipped in spilled milk and remnants of Gaucho peanut butter cookies all over the bathroom floor, you knew the Incredible Invisible Boy had struck again. We never denied him this invisibility and would sit at the kitchen table and watch him negotiate a complicated refrigerator caper with his eyes clamped shut in concentration. He was an abnormally cagy five-year-old, but my mother finally squelched Brian's superpowers when his invisible dance atop an end table sent him to the emergency room for twenty-two stitches. As a slightly older child, I still envied his perfect illusion of invisibility, and as an adult I wished I could tap into that kind of disappearance more times than I care to mention.

And on towers sometimes you got close.

There can be hours of downtime while you are up on a tower, when your work cannot proceed until some vital function has been completed below. If you went down, you'd just sit and then have to climb back up, and rather than do that, a lot of us would just hunker down and wait. From that perch you could get a real sense of the neighborhood you were in, a perspective people below could never appreciate. You could uncover secrets. You could eavesdrop and Tom-Peep. You could gauge the success of businesses. You could assess the health of relationships. You could clock the daily routines and the basic rhythms of life down there.

And you could fuck with it.

Crackhouse Tower was located in the Mechanicsville area of Atlanta, south of the state capitol building and west of the old Braves stadium, a part of town skipped over for any type of renewal while Atlanta was doubling and tripling its population in the outer suburbs. Interstates, railroad tracks, and outlying industrial parks hemmed in the neighborhood, which at our site was comprised of single-family shotgun shacks and several sprawling public-housing units. To the Metro Police Dispatch Unit it was known as Crime Zone 3. Not the worst place in the city, but far from safe. The section was a wreck at the time, the kind of place where when one of the locals would wander up and say, *You don't want to be around here at night,* you did not have to be told twice. I was atop the two hundred–foot self-supporter with Hangman, and Power was down below with Clay, one of the original Beatles. Below us was a narrow forgotten lane with six or so of those shotgun shacks, one of which was a thriving crack house with a heavy steel door. Two doors down from it was a clapboard Baptist church no bigger than eight or nine pews. Forty yards down the lane was what in New York we would call a *bodega* but here they called a *package store.* It was the only surviving business capping off a strip of long-boarded-up services and was girded in cast iron window bars and drop-down steel shutters. Behind it was a small yard wrapped in chain-link and razor wire and crammed with fifty-five-gallon drums of recyclables. It was early on a Sunday morning, and even though there was not a soul on the streets, I knew (having been at that vantage point for almost a week) how the day would unfold.

At six thirty the shutters of the package store would roll up. The first customers were always the stooped-over, worn-down grandmothers shopping for milk and bread. At the housing project, the dumpster divers would begin sifting through the trash for cans like punctual trappers checking their lines. They'd arrive at the store between seven and nine and exchange their plastic bags of aluminum for 40 Dogs wrapped in paper bags. Some of them would spend all day there bumming change until they had enough for another bottle. Some wandered back into the projects. One of them, always wearing a yellow do-rag, had his own umbrella for shade, and he would sit on a milk crate acting as one of two lookouts for the crack house. The other lookout sat on the porch of the house. He was also the cashier. A car would drive up, and he'd take their money and their order, place it through a slot in the big steel door, and deliver the goods. They were as efficient as a McDonald's drive-through. The steel door was absurd because the house was so ramshackle and rotten that if you wanted to get inside, you could just push through the walls. By ten or eleven, business really picked up because the addicts in the housing project were stirring into consciousness and needed to score. And from every one of the twenty buildings in that development, a path was beaten like *wheels over Indian trails* through the red Georgia clay, directly to the crack house. The preacher at the Baptist church would arrive every morning at around eight thirty and spend his day working in the yard, cleaning up the dross from the night before. He would stand defiantly at the corner of his lot and rain hellfire and damnation upon the junkies two doors down, who were sometimes lined up three or four cars deep.

"Don't do me one damn bit of good," he told Power a few days earlier, after informing him, *You don't want to be around here at night*. "They just laugh at me. I gave up on the cops."

The policing of Crime Zone 3 was cursory. The cops were not unsympathetic but understaffed and overwhelmed. This town was growing so fast they could barely put up traffic signals, and many beat cops had been relegated to directing traffic at the countless new intersections in the Northland. At first they responded to many of the preacher's calls, but by the time they got within two blocks of the scene, the lookout at the package store would wave his yellow bandanna at the lookout on the porch, and that was that. There was no longer any crime to be investigated.

"Just *all* be that way now," said the preacher. "You can't put the fear of God in peoples that's already dead. I just wish on Sunday they'd keep the faith for a minute or two. Used to fill this place up. Every week the pews gets emptier. In my salad days I'd go at 'em with a bat, but they don't think nothing of shooting nobody. Used to be a *nice* neighborhood."

Power and Clay were having enough trouble shooing away the constant annoyance of the *curious*, visitors like the skells up in Bayonne, whose eyes never met yours but rather darted around your job trailer and the reels of coax, their crank-addled minds calculating the value, which was always pennies on the dollar by the time they sold it. If not for this, I know Power would have done *something* to dissuade the crack house traffic for at least one Sunday. He was that way. He was the guy who *would* stop on the highway to help some stranger change a tire or troubleshoot their engine. He was the guy who would try to find out *Whose dog is this?* when he found a stray rummaging around the motel dumpsters. If he could find a skell he thought he could trust, he would toss him a few bucks for watching the site, even though he would remove anything of value when we rolled out at night. He, like the cops, felt there was really nothing he could do when the crack trade increased just as the early parishioners started to arrive at the little church, dressed to the nines. But he had not factored in the power of invisibility. And Hangman and I had.

That morning we brought a shiny new whistle up the tower, and every time that cashier stuck his head inside a customer's car window, we would give it a blast, and they'd scatter like rats. We thought we might get away with this once or twice, but they never caught on. The customer would speed off; the lookouts would shrug at each other in fear and confusion. Every time we blew that whistle, I could see Power drop his head and I'd watch his whole body shake, laughing his pony-tailed ass off. He gave us the thumbs-up and the *keep it going* hand signal. After about an hour, the cars stopped coming, both lookouts cowered inside the crack house (peeping out the windows every now and then), and the Sunday meeting was conducted in peace. After services, the congregation stood on the front lawn as if they had never seen it before and chatted with the preacher for a long time, something they had not done in months. When the church lot emptied, the dealers and their lookouts held a little conference outside the package store, and in a little while it was back to business as usual. Hangman and I decided not to intercede. After all, the reason we were on the tower was to replace all the coaxial feed lines that had been riddled with bullets.

But they never saw us. *They never do.*

Besides being invisible, we could also make things appear out of nowhere. In terms of construction, it does not take that long to erect a tower. Once the foundation has been laid (sometimes months before the stacking crew shows up), if you have the right crew and the right weather, you can put up a two hundred– to five hundred–foot tower in a few days, a monopole in a matter of hours. One morning in Atlanta, we erected a 180-foot monopole out by Six Flags, just south of many small municipal airports. But nobody told the pilots who left that morning that when they came *back* there would be a tower directly in the flight path of one of their approaches from the south. We could hear the small jets gliding in, then gunning their engines in evasive maneuvers. Powering up and left. Up and right. Up and over. Brody was halfway up the tower tightening step pegs, waving *WHAT THE HELL?!* to all of us on the ground. The other top hand said, "I could look into the goddamn cockpit! I could see the fear in their eyes!"

Brody said, "Maybe we ought to hook up a light."

Our invisibility was not limited to sight lines. On towers the airwaves themselves presented many opportunities for unwarranted and unwelcome manipulation. Our handheld radio sets were constantly picking up peripheral traffic. Fast-food drive-throughs were a common annoyance. On a particularly difficult stack in the southeast-Missouri tundra, our communications between the top hands and the crane operator (or "op") were constantly interrupted by disembodied voices ordering tacos and burritos all day long. There we were, Bo, Gunn, and I, out of clear sight of the op, feeding him very precise instructions while inching the big steel flanges into place, and then all the op hears is, *You want that tea sweetened or unsweetened?* The op did the right thing and just stopped, but he did so abruptly, and the whole section bounced and swung over to me, knocking me off my perch and sending my tools and hard hat *kerplinking* and *kerplunking* off the 240 feet of steel we had already stacked. I yelled HEADACHE,[62] and the groundhogs scrambled for cover. My spud wrench missed the windshield of the crane by a few feet and completely imbedded itself in the hard frozen dirt below. I checked myself to see if there was any real damage and scrambled back to position, cursing the op the whole time.

"Don't blame him," Bo said. "That's *your* fault. You didn't leave yourself an out." He was right, but I said fuck you anyway, and Bo taxed me two six-packs for dropping my tools. I took my revenge out on Taco Bell, getting on my radio and ordering about thirty bucks worth of quesadillas from a car that somehow disappeared between the drive-thru and the cashier's window. The Federal Communications Commission would not have appreciated that or my language to the op. The airwaves have rules, and we have been warned off many channels for unknowingly breaking them.

> *Regulation of Obscenity, Indecency and Profanity: It is a violation of federal law to air obscene programming at any time. According to Title 10, Section 1464, whoever utters any obscene, indecent, or profane language by means of radio communication shall be fined under this title or imprisoned not more than two years, or both.*[63]

---

62. *HEADACHE!*: What tower dogs yell when tools, equipment, or hardware is accidentally dropped from a tower to alert those on the ground. If sufficient cover is not close by, standard procedure is to stand fully erect with your arms at your sides, make yourself the smallest target possible, pucker your asshole, and hope for the best.
63. FCC, "Regulation of Obscenity, Indecency and Profanity." www.fcc.gov/general/regulation-obscenity-indecency-and-profanity.

Swearing over the radio is a fineable federal offense, though I have no idea how it could possibly be enforced. If it were enforceable, most of our paychecks would go directly to the FCC. We have no seven-second delay or dump button, and if it weren't for profanity, most of us couldn't complete a sentence. One crew leader got on the radio and laced into his top hand with such a creative strain of curses he was called on the carpet not by the FCC, but by the principal at a nearby grade school. It seems that right after the Pledge of Allegiance the kids there were treated to a profusion of *cocksuckers*, *motherfuckers*, *assholes*, *sonofabitches*, and *dickheads* spewing out over the school's PA system.

"I do *NOT* know who you are," the tower crew heard over their handsets, "but right now I am staring at a group of very wide-eyed second-graders."

These were radio frequencies and not cellular communications, but in the '90s (and to a lesser extent today), cellular calls could indeed bleed over into our two-way radio communications. Late one tepid summer evening while 450 feet above the Mississippi Yazoo Delta, I was made involuntarily privy to the saddest conversation I had ever overheard. Actually, it was half a conversation. I could only hear one side of it. The speaker's name was Luanne.

"Gretchen," Luanne said, "I just want to die."

It was a tiny voice. Deeply southern. And she was mannerly, stopping herself halfway through each swear word.

"I just want to f . . . *froggin*' die."

It was exactly 10:15 P.M. And I had been on the tower since seven thirty that morning. It was one of those sites we just could not get done. Spotlights had been set up in the compound so that we could work through the night.

"I just don't know what to do," she said.

Her melancholy suited the terrain, the steaming cotton and rice flats east of the Big Muddy, clouded with voracious mosquitoes and eye-sticking gnats and the stench of rotting vegetation. Some people were still living in tar paper shacks down there. Some were still fishing for insecticide- and pesticide- and fertilizer-laced catfish for sustenance. Most were living just barely above the poverty level while sitting upon the richest soil in the world. By the light of a harvest moon, the miles of fertile loam spread out like pounded lead. *If I lived here I would want to just froggin' die, too*, I thought.

But my glibness subsided as I listened to the depth of her dilemma, the gist of which was that her froggin' man had not only beat the froggin' shit out of her but went to South Carolina with both kids ostensibly to visit his mother, but neither his mother nor anyone else had heard from him since. He had the kids. He had the '93 Ford Taurus. He had the Walmart MoneyCard, which he bottomed out shortly after he left. She had the apartment that was four months behind in the rent. She had the bruises below her right eye and above her right breast.

"He punched me in my f—froggin' teat," she said.

Teat? Who talks like this? I did not know, but I wanted to know. She had the longing in her heart for her children, but most incredible of all, she missed *him*. That is when I got on the line. After listening to Luanne say the same things three nights in a row at exactly ten fifteen each night, I could not take it anymore.

"You are *beautiful*," I said. "You don't need this shit. You don't need this asshole in your life."

"WHAT? Who is this?" she said.

"This is your guardian angel."

"Who is with you, Gretchen?"

Nobody, Gretchen apparently said.

"This is your guardian angel," I said. "If you do what I say, your world will be all cotton candy and circus elephants."

"This isn't funny. Who *is* that, Gretchen?"

It took a while, but Gretchen finally convinced Luanne she was the only one on the line. That whoever Luanne was hearing was not with her. And Luanne then went from suspicious to curious.

"Who are you? *REALLY*," she said. "You're scaring me."

"I am the *Night Whisperer*. I scour the airwaves in search of those in need, dispensing unsolicited advice to the downtrodden and hopeless. Look out your window. Look up in the sky! See those red flashing lights blinking against the blackness? Any one of them could be me. I am the *Night Whisperer*, and I am everywhere."

To my amazement, she did not hang up.

"Well, gracious me," she said, and she laughed.

"See?" I said. "Don't you feel better already?"

I could *hear* her smiling, and then the connection fritzed off. The next night at ten fifteen, she crackled into my radio right on schedule.

"Are you there?" she said, a guarded hopefulness in her tone.

Gretchen replied in the affirmative.

"Not *you*," said Luanne.

"I'm here," I said.

"Gretchen hung up," Luanne told me. "She says I've lost my mind."

"Gretchen has no imagination."

"Am I imagining you?"

"What do you think?"

"Well, you ain't no guardian angel, I *know* that. But you're nice enough."

"I don't think you know what nice is anymore," I said. I then proceeded to tell her how to fix her life. And to my amazement, she listened, though my advice was little more than she could get at the local Family Services office and was also shit she had heard a hundred times before from people like Gretchen, but coming from the *Night Whisperer*, it may have carried a little more weight: *Go to the cops, report the bastard, find out what your rights are, proceed accordingly. Put up a fight, dammit; don't just sit there wallowing in victimhood.* It is easy to dispense tough love to a stranger. If nothing else, I prodded her out of her self-loathing for a few minutes.

I didn't hear from her for several nights, and when we stopped in the nearby town of Glen Allan (population: 681), I'd eyeball the local women, deluding myself into thinking I could match that voice to a face, which proved to be difficult because I did not even know if Luanne was black or white. But it would have been nice to have met her. We could have gone to one of the famed delta backwater juke joints and listened to the blues all night while downing cold Dixie tall boys and slurping up oysters on the half shell with lemon and Tabasco. We could have been friends. We could have been lovers. But I was on that tower for three more nights, and at ten fifteen each of those nights, the only thing my radio crackled out was static. A lonely, rueful static.

I thought about Luanne and about invisibility as the jet ferry rocketed toward Martha's Vineyard because the man I was going to meet had also come to the Vineyard to be invisible. That first ferry out of New Bedford was the 9:00 A.M. *Seastreak*, which completely lived up to her name. At first she crept at a leisurely pace from State Pier and then picked up a little momentum across Buzzards Bay and through the Woods Hole gap, but once she slid through the gap in the big rock jetty, she amped up to forty-two knots and burned water. This vessel held five hundred people and crew and still managed to shoot out rooster tails twice the length of the ship and three times as tall. The droning hum of it all was like a long-needed massage. This maniacal marine machine pulsated throughout my entire body, but benignly so, almost maternally. It was the hum of the womb. Being a Sunday and so late in the season, the ferry had few customers, so there was plenty of room at the full-service bar, where I had a Bloody Mary and slumbered into the persistent diesel droning of those engines, which comes a close second to rain on the roof for the chronically sleep deprived.

Marc Shmuger, his normally well-trimmed goatee looking errant and happily so, met me at the Vineyard Haven landing in shorts, sandals, and a white collared shirt. We drove around the pond to the house he had rented on the northeast shore of the island. The driveway was long and shouldered in conifers, and we passed the tennis courts and the pool and the volleyball court before we crunched into the gravel at the main house. It was a rambling place, at least one hundred years old, all wood and dampness and yellowing linoleum and overstuffed sofas and too many throw pillows, and all of it shaked in that cedar siding and roofing that turns pale gray just a few months after it's installed. It had the musk of a seldom-used cottage, and that was part of the appeal, I guess. It was right on the water, perched atop a bluff, and leading down from that a rickety and barely maintained oaken and rope stairway rambled right to the shore. There sat a small two-room cabin only ten feet from the surf. This was my accommodation. Marc apologized that I would have to stay there because he had scheduled more guests than usual. I told him he was out of his mind for being worried about it because this place was *perfect*. It had a bathroom and a bedroom and a living room with an ancient grilled gas burner. It was pure solace, and I was thrilled.

"Get settled," Marc said, "then we can talk."

Up at the big house an hour later, we sat at the long slab table in the kitchen. He was there with his wife and two kids, whom I had met in L.A. years earlier, and also a renowned cinematographer and his wife. Lovely people, all, who did not want to talk about what I wanted to talk about, which was movies and how the fuck did I get them to hire me to write one. I wanted to write the tower dog movie. I wanted him to pay me to write the tower dog movie. But I couldn't pitch it. This was, after all, a social visit. I was there to say hi, old friend, and to cook a nice meal on the morrow and to catch up on old times, but we both knew at some point I was going to get to business. What we did not know was when. This was, after all, his vacation, but judging by the piles of Fed-Ex envelopes from Universal Studios in the living room, it was, at least for him, a working vacation. How did I tread this ground? The best plan was not to tread at all.

Dinner that night was to be supplied by two of Marc's other acquaintances who were presently fishing out somewhere between Vineyard and Nantucket sounds for stripers and blues. Marc got a cell phone call and started firing up his grill. The boys were coming in loaded. I asked the cinematographer's wife who was bringing in the catch, and she said, "A writer friend of ours," and that set me to pouting.

As I mentioned earlier, I cannot stand being with writers. We are the most self-involved, self-important assholes you will ever meet. Especially the bad ones, in whose company I might belong. We think our take on the human condition supersedes your own because we took the time to scribble it down. And that hubris is and will always be our defining delusion. We don't want to work. We want to write. We want to act. We want to sing. We want to do all the things that garner attention without actually doing, to us, what is labor. So hearing that another writer was here on the Vineyard to visit chafed my ass. I grew more and more petulant as I sat on the back deck and sipped a beer and watched the Boston Whaler pull up to the shoals below and the writer disembark with a fat stringer full of striped bass and bluefish. I hadn't even noticed there were more people arriving for dinner. The writer-fisherman had a young daughter, and she emerged with Marc's two kids, and the copilot on the Whaler showed up, and soon the backyard and porch of the house was vibrant with the laughter and high-pitched twiddling of kids and the back-slapping familiarity of what I perceived to be some sort of secret Hollywood society that did not include myself and the smoky goodness of stripers and blues that just an hour earlier were streaming through the deep blue—and there sat I, inwardly detached.

The writer grabbed a plate of striper filet that Marc had grilled to perfection, the white meat steaming and flaky and the black-and-silver-speckled skin crisp and fatty and curling deep brown at the edges. I opted for the blue, a roanish dirty meat with a slate-black skin and blue-collar rep. The writer sat next to me on the steps of the porch. His dark wavy hair was whitened with the ocean, and he said, "Hey, I'm Paul."

And I said, "Hey, I'm Doug."

He said, "I hear you're pretty good."

I thought, *High praise, thank you, coming from whom?*

"Excuse me," he said, and he wiped his hands on his shorts. "I'm Paul Attanasio," and as he shook my hand, I leaned back in disbelief and almost knocked his plate of fish off the deck.

Heroes. I have a few. My father, of course, but nothing could be more subjective. Terry Bradshaw broke my thumb at Offense-Defense Football Camp in Vermont one summer. In a meeting in Harry Gittes's office in L.A., Jack Nicholson asked me, "So do you want to act or direct?" and I said, "No, just write," and he said, "*Thank fucking God.*" I had met Ron Swoboda, who made one amazing catch in the 1969 World Series. And I had one of the greatest poets of my generation tell me I sucked. These were heroes I actually met, and I appointed myself as unmemorably as all get out. The other heroes you know you will *never* meet and you resign yourself to that . . . George S. Patton, John Lennon, Jimmy Carter, Ian Anderson, Paul Newman, Paddy Chayefsky, John Steinbeck, Aaron Sorkin, Derrick Thomas, and Paul Attanasio. I was gobsmacked. As I steadied his plate of fish, I began rambling about his work in television and film and my absolute love of *Homicide: Life on the Street,* and he smiled politely and got another plate of fish and said, "You know, we never found our audience for that one," and that absolutely blew my mind.

When is success not success? When is critical acclaim and financial reward not success? What do you have to do in this frigging business to be satisfied? As the embers in the grill glowed from red to white and the kids got quieter and the dusk descended, I realized that this man was born in the Bronx and I was born in Brooklyn and that we were two weeks apart in age and that he had the professional resume and career that I only could have dreamed of and above all he was this—dissatisfied. He wanted to do better work in a world where acceptable was okay. *Quiz Show, Donnie Brasco, Homicide.* He humbled me to the point I could not even think about pitching Shmuger on anything, and he didn't even mean to do it. I became invisible again, a deceptively simple talent, and I went back down the rope-stairs to the little cottage on the shore and nestled into the quilts on the camelback mattress and closed my eyes to the slurping of the gentle waves and the knowledge I was totally fucked.

The next morning, I sat at the bottom of the steps sneaking a cigarette, which I was only doing because Marc disapproved of smoking. He caught me like a sophomore in the bathroom during homeroom hour. He ignored that fact and sat next to me and asked, point blank, "What is your favorite movie of all time? I mean, what do you think was the best movie ever made?"

Was this a test? I didn't have one favorite movie; I had a hundred. They scrolled through my head like a clattering, blinking slide show: *Citizen Kane, The French Connection, Network, The Longest Day, It's a Wonderful Life, The Ox-Bow Incident, The Sting.* But what I came up with was . . .

"*Cool Hand Luke,*" I said.

He rubbed his newly scruffy chin and stared at the sky. I remember him doing that at our meetings at Columbia. He'd get quiet, then cock his head slightly back and to the right, and stare at the fluorescents and rub his chin and just think, breathing in and out heavily. Then he would exhale one final time and finally speak.

"*The Godfather,*" he said, and of course he was right.

Later that morning we went to town, and Marc and the kids went for haircuts and the kids got Mohawks and came out of the barbershop giggly as hell. This part of the Vineyard reminded me of any one of a dozen Long Island East End towns and hamlets. Faded and warped wood, rope garnishing, hanging buoys, and lobster pots. Cracked asphalt lanes half covered in sand. This was old money making tons of money off of new money and resenting them for it. And there was no absence of "Locust Valley Lockjaw."[64] Mrs. Shmuger and I went to the fish market to get the clams and mussels and shrimp that I was going to cook that evening. Of course, she tried to pay for it, but to my credit, and as broke as I was, I did not allow her to pay for *most* of it. That night we sat at that big slate table in the musty old kitchen, and I blew the whole meal. The clam sauce was fine, but I ruined the pasta. Overcooked the shit out of it. Then I doused the mussels in way too much butter and too little white wine. Marc and his wife and the two kids and the cinematographer and his wife ate most of it. They were phenomenally civil in their disappointment. You could not tell I screwed up, which is the way L.A. works anyway. Marc took me to the landing the next morning, and as we sat waiting for that mother of a rocket boat to steam in, he put his hand on my shoulder, and he said, "I think the tower thing is TV."

"You think?" I said, earnestly, having never thought of that.

He rubbed his chin and stared at the slatted wooden overhang and breathed in too much. He finally exhaled and said, "Yeah. I really think so. Call me next time you're in L.A."

---

64. Locust Valley, an old-moneyed unincorporated area on the north shore of Long Island, is famous for their Thurston Howell III–fashion of talking without actually opening their mouths; hence the term "Locust Valley Lockjaw."

AUGUST 31, 2015

A fall has claimed the life of another tower climber. Police say a man fell to his death while fixing an antenna atop a water tower in Brownsville, Texas. Authorities responded to the scene and found 21-year-old Pedro Samuel Salazar on the ground.

Salazar, who was an employee of Orbit Broadband, was transported to a local hospital, but later died from his injuries. Witnesses said Salazar lost his grip and fell about fifty feet after trying to perform routine maintenance on a satellite antenna atop the water tower.

"We at Orbit are deeply sorrowed by this and our prayers go out to the individual and his family," Richard Galvin, broadband operating director at Orbit, told a local TV station. "There is an ongoing investigation and we are working with all those involved. Out of respect for his family and his coworkers, we ask for some privacy during this very difficult time."

Investigators have ordered an autopsy in the case to determine what happened.[65]

---

65. *RCR Wireless News,* "Another Death in the Tower Industry," August 31, 2015. www.rcrwireless.com/20150903/cell-tower-news/cell-tower-news-another-climber-dies.

CHAPTER FOURTEEN MISSISSIPPI
A LITTLE DOSE OF DIGGER

In my experience there were no broad strokes with Shmuger. There were no *run it up the flagpole* shots in the dark. Everything he did was specific and thought out, and I knew he would not tell me what he was thinking until he was sure of the exact way to proceed. He was a lot like the Godfather (both the Marlon Brando and the Jimmy Tanner versions) in his clinical assessment of the angles. So I knew that when I did go to L.A. and make that call, that ducks beyond my knowledge would already be in a row. And they were. Fallon and I went to Azusa. We had the meeting. We hit that home run. And over the next few days, amid the reeling synchronicity of my past colliding with my present, my future was secured. I parried with the entertainment industry in an attempt to bring national media attention to the plight of the tower dog. Even though I was out of my depth, it actually worked. And why wouldn't it? After all, I had promised them death. And though the talks were still progressing, and no contracts had been signed, I did what I had always done when the grass on the other side of the tower dog fence got tantalizingly greener. I quit K.M.C.A.

I went home to my tired and leaning house in the country about six miles southeast of Lawrence, and I cradled my baby and his mother, and I cooked and I cleaned and I sat on the porch and fired up the chiminea. I was home . . . *HOME*. Every evening as the Kansas coyotes wailed in their netherworld between dusk and dark, I would open all the upstairs windows and let the brisk fall wind blast through the house and sit behind my office desk and place everything just right—mouse, keyboard, monitor, stapler, notepads, highlighters, felt-tips, sticky notes (staging for the juggernaut of creativity to come)—and wait for the show to begin. *Tell Jungle Boy Delaniac is once again a writer.*

Though originally entertaining the thought of going directly into production with a reality show, the network decided to hedge their bets. They decided instead to produce a *Dateline* special, which would be one of the first of its kind for the long-playing news magazine: they wanted to produce a one-hour tower dog show in the hopes of subsequently marketing it as a reality series to the highest bidder. They were gonna run it up the flagpole to see who saluted. This little tidbit came to me through the offices of Ben Silverman, head of NBC Entertainment at that time. And I was instructed to contact David Corvo, executive producer at *Dateline*. Corvo, a product of Waterbury, Connecticut, and UC Berkeley, with over thirty years of broadcast journalism under his belt, was soft-spoken and articulate, affable and not surprisingly corporate. His plan was to send one of his "top producers" across the Hudson to Ramsey and meet the dogs in order to assess whether or not their plight was worth burning one of the hundred or so annual hours of *Dateline* episodes Corvo would exec-produce that year.

Sausages, laws, and network television. In one phone call I slipped from being at *least* a creator and coproducer on a network series to being auditioned in the parking lot of the Extended Stay motel for a possible one-off for a weekly news magazine. This left turn was disconcerting. But I agreed, even though my consent meant nothing without the cooperation of Brody Kilfoyle, who did not embrace the news. Both he and I wanted our world to be explained, explored, and celebrated, not scrutinized. We wanted to show the world just how hard the job was, just how capable these men were, and just how much they sacrificed so that America could simply operate on a minute-by-minute basis. We wanted to show the world how what we did was important to *everyone*, right now, right here. Brody's reservations were many: the safety of his men while filming, his distrust of media in general, his concern about film crews interfering with the work at hand, his worry about his business relationships with major general contractors and carriers. All of these were legitimate concerns. Plus Big Jim, Power, several of K.M.C.A.'s field foremen, and, I am sure, Brody's wife, Katt, were against it as well. None of us wanted this *Dateline* hour. And there was no damn money in it. Not for me, Brody, or the dogs. The prize was the series, and *that* I did want. This was the carrot I had sniffed in NYC so many years ago. I would be in the biz for real. Not on the fringes, not a bone here or there; I'd be in mainstream television. Brody and I kicked around the pros and cons for days before coming to the conclusion that we would meet with this "top producer" and give him as much of an audition as he was giving us. If the vibes were choppy, we could bail. And we also, perhaps naively so, came to the conclusion that *fuck it, they can only show what we allow them to show. We might be all bozos on this bus, but we are driving the bus. Alpha males, all—we can control this.*

The "top producer" was currently elbows deep in the Natalee Holloway[66] disappearance, splitting his time between 30 Rock, NBC's L.A. studios, and the field, but *Dateline* had apparently felt strongly enough about the tower dogs that they arranged for both of us to fly into Newark and then drive out to Ramsey. That producer, Tim Uehlinger, picked me up outside the terminal in a rented convertible.

66. Natalee Holloway was an American teenager who made international news after she vanished on May 30, 2005, while on a high school graduation trip to Aruba, a Dutch island in the Caribbean.

He wore a light blue collared shirt unbuttoned at the neck, khakis, and a loose, wrinkled blazer. He had a strong jaw and bright eyes and a thin, contemplative grin. He had long ago lost whatever New York accent he might have had, a good thing for a broadcast journalist. We headed north, and I discovered Tim and I grew up about seven miles from each other on Long Island. We were a year apart in age. We had a mutual friend, a man I played youth football with who was a classmate of Tim's at Chaminade High School. Tim had been rerouted from a major international story to meet a bunch of roughnecks, and if he resented that, you could not tell.

"I can't promise you anything," he said.

"I know," I said.

"This is just a meet and greet," he said.

"I know," I said.

"I don't usually do this kind of thing," he said.

Outwardly, he was slight of build and professorial, a self-described "milquetoast" (his word), but that I would learn was far from true. He was as wiry and tough as some of our climbers, but he did not know that because the energy that almost frenetically pulsed through him was mostly spent in his head and on the dozens of stories he was juggling that had become his life. He literally had a spring in his step, each stride not only sending him forward but somehow *upward*. He was a newshound. And when he dwelled on something he had the habit of spinning his finger about one spot on his hair like a fork spinning spaghetti. In the year that followed, we would grow to hate and love and ultimately respect each other. But driving up Rte. 17, it was as uncomfortable as a blind date. *Who will be the top dog here? Do we want the same things? Does he drink? Does he smoke? Will we endear him to us or repulse him?* It was fifty-fifty, but at least I'd know in a few hours.

We arrived at the Extended Stay mid-afternoon, and the troops were all out in the field. Tim and I checked in (he had a small overnight bag and a laptop), and only then did I take the time to get online and find out a bit more about him:

NBC News senior national producer Tim Uehlinger has covered the world for NBC News for nearly a quarter century. Tim produced much of the Today *show's pioneering eight-day trip to Africa in 1992 and was among those honored by UNICEF ambassador Harry Belafonte and the United Nations for his coverage . . . [He covered] the first Gulf War and the invasion of Kuwait by Iraqi forces in 1990, and the Somalia relief efforts in 1992, the Oklahoma City bombing in 1995, the death of Princess Diana in 1997, the September 11 terror attacks in 2001, the first television interview with Pakistani rights activist Mukhtaran Bibi in 2002, the Iraq war in 2003, the Indonesian tsunami in 2004 . . . Tim has been on scene for most of the major news events and humanitarian crises of our times. He is a multiple Emmy Award winner. Most of Tim's career at NBC News has been spent at* Dateline NBC, *where he has produced long-form breaking news specials and a wide variety of hour-long features and investigations.*[67]

Damn. I could now officially remove the quotation marks from "top producer." *Dateline* had rolled out one of the big guns.

It was prearranged that the Three Wide Men and I would have that meet and greet in the parking lot. After the trucks rolled in and the men had showered and changed, they assembled in twos and threes by the pop-ups until about a dozen or so were grilling burgers and dogs and popping beers and passing the fruit jar. I would not say they were subdued, but they were not as characteristically rambunctious. After all, it was early. I introduced Uehlinger to Sarge and let go from there. I did not shadow him. The boys would be the boys. We couldn't control their off-site behavior in the past, and we didn't expect to now. But I did notice that most of the Children of Andrew and other morons were mercifully absent. Whether or not

---

67. LinkedIn, Tim Uehlinger's page. www.linkedin.com/in/tim-uehlinger-a984b87.

they were transferred from market or given the dog I couldn't say, but some serious house cleaning had been done.

I settled into a folding chair on the fringes and watched as Tim moved from pocket of men to pocket of men, handshakes and small talk. He had a burger, had a beer, had a cigarette. That was encouraging. After a while, they moved toward him. Soon I heard laughing and snippets of the same old war stories that had become our parking lot stock-in-trade. Then something happened that made me feel we were going to be okay, because I realized Tim wasn't selling Tim. Tim wasn't selling anything. He was asking them about *their* lives: *Where are you from? You got any kids? How long have you been doing this? What does your mother think about your job? How did you get THAT godawful nickname?* This was something nobody was used to. This guy is genuinely interested in US? *Sonofagun.*

By nine or ten, it was over. Tim excused himself and went to his room, where he and I chatted for a while. We talked about some of Brody's concerns, about scrutiny. How we wanted this to be informative and not investigative, to be entertaining and not muckraking. Tim agreed and allayed some of that angst by schooling me on a few things I frankly was not aware of, the main point being that *Dateline* didn't just do investigative sex, scandal, and murder-of-the-week stories, but often did "softer" stories, features, slice-of-life chronicles on one subject or another. That is where he felt the tower dogs project fit into the schedule. That was palatable, and I relayed this to Brody in Nashville. We both breathed a little easier. Though Tim was playing his cards close to his chest, I definitely sensed he was leaning toward the green light. Yet, still, what had been crystalline in my head just a few weeks earlier had become blurry. *What exactly am I doing here?* Ethically, what NBC proposed was whistle clean. There was precedent. But conceptually the project was mired in gray. No matter what we called this thing (a special report, a feature, a slice of life), anything presented under the banner of *Dateline* had to be considered journalism by association.

My experience with journalism was minimal; for about two years I wrote fluff and covered sports for the Southwestern College *Collegian* and the Winfield *Daily Courier.* How do you sell papers? Mention as many names as possible and try to spell them correctly. I fancied myself the Jimmy Breslin of Cowley County, population eleven thousand. The roving reporter, covering girls' basketball and amputees, chamber-of-commerce ribbon cuttings and track meets. That was so long ago I had forgotten about it until that night in Ramsey. And of the dozens of articles I tapped out, only two of them came to mind. In one I called a college professor "the human sedative." I got called on the carpet for that one, and though I staunchly defended the right to free speech over accusations of libel and slander, I still felt shitty about being that much of a smart-ass in the first place. I was going for the laugh, and the laugh wasn't even mine. I stole the line "human sedative" from a classmate, Wade Cargile— a sort of plagiarism, if you get technical. The second article was about Max Thompson, a renowned ornithologist who discovered a new species of duck, the Thompson steamer duck,[68] not a small accomplishment when the entire bird world wrongly assumed there wasn't a duck on earth that hadn't been classified. What got me in trouble there was not trying to be a smart-ass but my own sincere incredulity—because part of verifying the existence of this duck was to shoot it, which Max did. And I made a point of that. And I punted the science. The Thompson steamer duck was not an *endangered* duck, just undiscovered as a separate heretofore unrecognized species of duck. There were lots of the damn things. I got chewed out for that by Max himself, God bless him. He had explained the science to me when I interviewed him, but I fixated on the wrong angle and played it out. There was no malice on my part, just plain sloppiness. And those recollections, that bitchy

---

68. "Their (Max C. Thompson and Philip S. Humphrey) examination ended in a scientist's dream come true: the discovery of a previously unknown animal species. In this case it was a flightless steamer duck, the first new species of the duck, goose, or swan family to be reported since 1917." *New York Times*, February 16, 1982.

clarity of distance, should have prompted me right then and there to take a page from my own history and learn from it instead of being doomed to repeat it . . . I should have known to walk away. Because what I had done was exploit a taciturn professor for yucks and foist my slanted point of view upon the irrefutably scientific. I hacked.

Was I, in collusion with *Dateline*, about to exploit the tower dogs for personal gain and profit? I was. Would the increasingly fuzzy lines between news and entertainment and reality television (which got fuzzier and fuzzier on a weekly basis as the "real" in reality programming grew almost laughably "produced") be crossed and crisscrossed as we stumbled forward? They would. And would there be unseen consequences? Hell, yes, by cracky . . . as invisible to me as Hangman and I were above Crackhouse Tower.

I checked in at home and then sat outside by the pop-ups and fished around a cooler for a beer. Lake Scum had dried up, leaving behind a slick layer of greenish glop peppered with crushed beer cans and flattened cigarette packs, plastic bags and dead birds. The dogs were elsewhere. After a few minutes, Devil Anse came around the corner, returning from the Olive Garden, trailing a few dogs behind him, whom I knew he had just treated to drinks and appetizers. They went inside, and Devil Anse came up and sat next to me.

"TV, huh?" he said.

"Looks like," I said.

"I don't think I want nothing to do with it," he said.

"Copy that," I said.

He clapped his hand on my shoulder and went to bed. It was barely ten, and I needed to get out of my head for a minute. I needed a dose of Digger. I called up Laurel in Franklin Lakes, and she agreed to slip out for a bit. We met at Grady's At The Station. Most of the commuters had headed home, and it was quiet. We sat outside, her with a cosmopolitan and me with a "two horses." No trains happened by. I have a habit of tracing one finger on my palm when I can't find words, and of course Digger noticed that.

"What's going on?" she said.

I laid out the whole deal, not my concerns or my confusion but how it was going to all be cotton candy and circus elephants. She let out a half-chuckle-half-*hmmmmm*, and I said, "What's that supposed to mean?" and she said, "Of all your projects . . . of all your plates in the air . . . what makes you think this one'll work?"

That made me laugh and shake my head. She laughed, too.

"*Riiiiiight?*" she said.

"Right," I said.

"We've been down this road before. *What makes you think . . .*"

The waitress brought us another round, and Digger graciously allowed me to pay for it. I suddenly felt exhausted. I raised my glass and shook my head and returned Digger's wise-ass grin and said, "Because it has to."

SEPTEMBER 25, 2015

The Hollidaysburg man who fell to his death on Friday morning was part of a five-person crew working on a radio tower.

Matthew J. Vance, 37, was employed as a tower hand for Com-Pros Inc.

He fell about ninety feet while working on a radio tower near Rote in southern Clinton County, Lamar Township Police Chief Martin Salinas said.

ComPros was hired to tear down an existing radio tower and build a new, stronger structure that can handle 911 emergency communications and other information.

Clinton County recently worked out a lease agreement with Vigilant Global Communications Co. for a new tower. Vigilant is paying the nearly three-hundred-thousand-dollar cost of the project.

Vance is survived by his wife, parents, stepfather, grandson, two stepsons, two sisters, two stepbrothers, and nieces and nephews.

His funeral service is set for 11:00 A.M. today at Evangelical Lutheran Church, Duncansville.[69]

---

69. *Altoona Mirror*, "Local man killed in fall from tower," September 25, 2015. www.altoona mirror.com/news/local-news/2015/09/local-man-killed-in-fall-from-tower/.

CHAPTER FIFTEEN MISSISSIPPI
OH, NIKE, YOU'RE SO FINE

The light went green less than a week later. It was decided shooting would begin after the upcoming Thanksgiving and Christmas holidays. But first Tim Uehlinger needed to discern exactly who and what he would be shooting. It was arranged for him and an assistant to come to the office-shop complex in Nashville, meet Brody face-to-face, and root around our world for a few days. At the time we had crews working in Jersey, Philadelphia, Memphis, Knoxville, Chattanooga, Atlanta, and Florida. There were also five or six crews either heading into, heading out of, or working in Nashville. By some odd luck these were guys both Brody and I would have loved for *Dateline* to follow. Jess Pulaski was there, fresh in from Ramsey, having let his bright orange hair grow down to his shoulders and his beard down his chest, consequently earning him the moniker Ginger Jesus. And there was Jack Kent and Danny Bush, young and proficient crew leaders. Not a lot of drama followed these guys. There were cursory introductions and not a lot else because even at the shop these guys were on the clock, and Tim honored his agreement to interfere with the job as little as possible. I spread the word that anyone who might be interested in participating in the program should meet us at a sports bar up in Goodlettsville about ten minutes dead north. Some declined but most did not, and that evening over a dozen dogs meandered in, every one of them a potential subject for the *Dateline* special. It was just your typical sports bar with too many TVs and drinks that came back at you with a receipt plus tax. Franchised and unimpressive. The dogs were not vying for stardom, but they were curious. And they were also something I'd never seen them be—shy. They shot pool, slung darts, had a few drinks, and noshed on the standard nachos and wings and mozzarella sticks. By eight o'clock they had gotten a little loud but no louder than the rest of the customers. They did know Tim and his assistant were there, but no one approached them directly. They knew they were being scanned by NBC for some kind of reality appeal, but they did not know what the parameters of that appeal was. And neither did I.

I sat with Tim and his assistant at the bar. He nursed a Michelob Ultra on ice, a combination new to me, and considered his prospects. He made inquiries. He was keen-eyed, almost predatory, studying, scanning, and twirling that fork of a finger in his hair. I got the feeling this was just some kind of gestation period he went through when handed a new assignment, a kind of internal struggle between *What the hell have I gotten myself into?* and *How can I make this work?* My quarters were up on a table, and as I was racking the balls, a dog leaned into me and whispered, "So who's the hot Mexican chick?"

The "hot Mexican chick" was Aliza Nadi, Tim Uehlinger's assistant, and she was far from Mexican. She was Afghan-American. In her mid-twenties, a graduate of the UC Berkeley Graduate School of Journalism. She had the deepest brown eyes I had ever seen, a mane of thick black hair just touching her shoulders, and a Mona Lisa smile that was as mysterious as it was infectious. Whereas Tim at this time comported himself with polite but professional detachment, it seemed that Aliza was thrilled to be there. She and her family lived in Glen Cove, Long Island, a place I had also lived in. We knew the same suburban streets and bottlenecks and LIRR schedules. There was an openness and enthusiasm in her I seldom encountered in my dealings with "industry" people. No affectation. No arrogance. She could jump from English to Pashto to Dari effortlessly, and to me she was just as bright as a penny.

We started to kick around some of our options regarding subjects. Tim had already wanted to visit with Cady and Daryn and Sarge and Devil Anse, still up in Ramsey, and they really liked Danny Bush, whose freckled-faced youth went against what central casting would label as a tower dog. They liked Ginger Jesus, but the problem was that he was the most monosyllabic human being I had ever met. He had a stock answer for almost any question you could ask him, which was, *I reckon.* Freezing up there today, wasn't it, Jess? *I reckon.* Going on eighty-five hours this week? *I reckon.* Did you just break three fingers, Jess? *Reckon I did.* It was cautiously decided they would follow as many crews as time and money would allow, and let the story lines reveal themselves. That seemed not only sensible, but fair, because in this room and even up in Ramsey, there wasn't a subject I thought would embarrass us. Frankly, aside from when the boredom of the road and the pressure of the work generated spasms of lunacy, we could be quite boring. That was perhaps a blessing for myself and Brody, but I knew it would be problematic for *Dateline* because any story needed a hook and a bit of sizzle. If the end game was that reality series, we would need a lot of *both.* And like an astonishingly implausible plot twist in a B movie, *BOTH* walked in the door at about nine fifteen in the form of Nike Rawlings and her crew.

I had never met Nike Rawlings and her crew. I had never even heard of them. I did not know they worked for K.M.C.A. because they were not an in-house crew. They were subcontractors out of Gadsden in far northeastern Alabama and only occasionally worked for Brody. They were a sub crew for us as much as we and many other companies were subcontractors to general contractors like GeoDyne Tech or Bechtel or Ericsson. They were no more direct employees of K.M.C.A. as we were for the carriers. And they were not at all shy. The first time I set eyes on them was also the first time Tim Uehlinger did; but where I saw *not good* he saw *hell yeah, baby*.

Nike Rawlings was an industry rarity because she was not only a female company owner and in-field crew leader in a predominantly male-oriented business, but she was also adorable: five foot one with bright blue eyes, sporting erratic blond hair and a sexy Alabama twang (that only she could make sexy). Nike had a killer body and a killer mind. To Uehlinger, Nike was as much the promised land as Rte. 17 was to Frogger. She was his hook. To the men who knew her, she was two things: a crew killer or a bitch. She was a crew killer when the men under her screwed up and she fired them on the spot, and she was a crew killer when the men above her pissed and moaned that she couldn't keep a crew. She was a bitch when she would not screw any man in either echelon. She ran her men as hard as Sarge and Devil Anse ran ours. Balls to the wall and unapologetically. She was a single mom whose preteen daughter was back in Gadsden in their quaint Victorian home under the care of her mother. As tough as she seemed, there was a sensitive, almost dainty side to her that only came out after dark, when the work was done and the hard hat came off. She was vulnerable in a world that pissed on vulnerability. There was no doubt in my mind that Nike stood a good chance of being dogged by cameras, and I did not think that was a bad idea at all. The worrisome component was her top hand, Ludlow Flagg, the self-described "*Dee*-troit City Cowboy."

Ludlow was big, bold, and brash. Over six feet tall and broad-shouldered, with short dirty blond hair and a thin Fu Manchu, he wore an earring, a puka shell–bead necklace, and a ubiquitous black cowboy hat that seemed too small for his head. When he found out why Tim was there, he strode up to him and said, "I am one badass tower worker. I've never found one better. I am the fucking best there ever was." That raised Tim's eyebrows a bit. This declaration I had heard countless times from countless men, but never from the real McCoys, only from the insecure and self-delusional. I did not know what kind of hand Ludlow was, but merely saying he was the best assured me he was probably far from it. Nike rolled her eyes and said, "When you bother to show up you might not be a half-bad hand." Ludlow cackled on about his prowess at elevation over a few games of pool to anyone who would listen. I couldn't read Tim's mind, but I did suspect the inevitable—that if *Dateline* wanted Nike, then by proxy they were going to get Ludlow. There sat the hook and the sizzle. And if (by the prevailing "wisdom" of 90 percent of reality television production then and now) you needed at least one asshole, then we were destined to have that, too.

The entrance of Nike and Ludlow into the mix, though ominous, did not curtail my enthusiasm for the project. I left Nashville energized and hopeful. I spent the entire holiday season at home for the first time in years, and it was glorious. On Christmas Eve my son turned one, and I was confident he would grow up in a household far more affluent than my own; Tim had offered me a production position on the upcoming shoot, and I would actually be making money. Weekly, it was about twice what I made as a tower dog, so who could bitch at that? And he did not *have* to do that. I surmised his reason for doing so was because I was his calling card unto the dogs and the industry as a whole. I was the icebreaker. But it was also because they needed someone to actually go *up* those towers with a camera, and that would be me.

Shooting began after the holidays, and true to his word, Tim did not put all his prospective subject eggs in one basket. From the first week in January until the first week of April, it seemed we went everywhere. We followed everybody. We followed Nike's crew in eastern Tennessee and the boys up in Jersey. We followed young Danny and his young bride not only on the job but also to church services where he sat in on the drums. We followed Bo and Gunn and Machu out in southeast Oklahoma. We did sit-downs with Brody and Sarge and the Godfather. We visited families of fallen workers in Kansas City and Springfield. We interviewed dogs in the air and bosses and managers on the ground and wives and daughters and sisters and brothers and friends of all. We wanted to shoot Daryn, but he had been arrested down on the Jersey Shore "for his own protection." This was his own damn fault. He and a few of the boys went to see some headbanger concert near the water, and something prompted Daryn to strip down to his undies and go for a dip in the ocean. This was in February. Some alarmed citizen called the authorities. When the shore police showed up, Daryn was about thirty yards out in the surf, and they shined their lights on him and asked, "Hey, you wouldn't be suicidal, would you?"

Daryn grinned that beautiful broad smile and said, "Maybe," and that probably wasn't the best answer.

California, Tennessee, New Jersey, Louisiana, Mississippi, Oklahoma, Pennsylvania, Missouri. I was flying so much that one time I missed a plane because I thought I was in Chattanooga when I was actually in Baton Rouge. And I *hate* flying. Let me modify that—I do not hate flying. I am not afraid of flying. I love flying. What I hate is airports and the people who work in them. There is a big difference. So I only flew if I absolutely had to. Otherwise the Explorer and I tacked up over seventy-five hundred miles, and NBC paid fifty-two cents for every one of those miles. What world had I landed in, and why hadn't I found it earlier? Then and now I have no idea what the budget is for a one-hour *Dateline* special, but it seemed bottomless. At every location we went to we were met by some sort of stringer team of cameraman and soundman for NBC or subbing for NBC. Aliza Nadi also always had a camera on her shoulder. I was fitted with a helmet cam and a handheld Sony camcorder for the aerial work. Tim, still working on multiple projects for *Dateline*, was sometimes gone for weeks. I'd get a call and be told *go there and shoot that,* and it was usually me and Aliza and that two-man stringer team on-site. So help me God I loved every minute of it.

In four months we had amassed hundreds of hours of footage. Tim even left the fold, so to speak, to follow a different crew from a different company out in California. I was jealous of this at first but then realized three things: that Tim was gonna do what he was gonna do, and I had nothing to say about it; that for sheer objectivity's sake, he had a responsibility to his story to check out other companies and workers; and that he was constantly analyzing and exploring in the attempt to drape the blood and guts and sinew atop what heretofore only existed as a skeleton of a story. He had miles of tape, over a dozen subjects, a hundred angles. If there could possibly be a deeper meaning in this, he probed for it. He justified money and miles. He agonized over where this story was going because frankly, after almost twelve weeks in the field and tens of thousands of dollars expended, this story was going nowhere. By some anomaly, though prior to this there had been one tower dog fatality every twenty-six days, January to April of 2008 turned out to be the safest quarter for climbers since the compilation of reliable statistics. But although we were going nowhere, getting there was one helluva ride.

OCTOBER 22, 2015

Ernie Jones, 65, a principal in Consolidated Engineering Inc. of Lynnville, Indiana, died Wednesday during an inspection of KOCO Oklahoma City's broadcast tower. Few details about the incident are available. Brent Hensley, president and general manager of the Hearst Television ABC affiliate, said the inspection was routine.

"Obviously, we are saddened by his death and our thoughts and prayers go out to Ernie's family, friends, and the people who knew him well, including our Hearst family," said Hensley. Jones's business associate David Davies said he had worked with Jones for fifty-two years at various companies. Incorporated in 1990, CEI has provided engineering analysis, tower renovation, and other services for a variety of broadcasters, including Hearst, CBS, Citadel Broadcasting, Clear Channel Communications, and Kentucky Educational Television.

The CEI website says Jones played an important role in the development of various tower- and structure-related standards, including the ANSI/TIA-1019 Gin Pole Standard. Jones also since 1986 had been a member of the telecommunications and electronics industry association responsible for writing and approving the ANSI/TIA/EIA-222 Standard, the American National Standard for Steel Antenna Towers and Antenna Supporting Structures.

Both Jones and Davies provided information for the Digital Tech Consulting study of the FCC TV band repack conducted for NAB.

Jay Adrick, an independent consultant and former Harris Broadcast VP, spent "a considerable amount of time" interviewing both Davies and Jones while working on the DTC repack report.

"Ernie was a very valuable resource to the industry, and obviously he will be missed, especially during this very important time of transition in the television industry," Adrick said.[70]

70. *TVNewsCheck*, "Tower Engineer Ernie Jones Dies in Acciden," October 22, 2015. www.tvnewscheck.com/article/89418/tower-engineer-ernie-jones-dies-in-accident.

CHAPTER SIXTEEN MISSISSIPPI
WE'RE LOST BUT WE'RE MAKING GOOD TIME

At a Red Roof Inn just south of Memphis, Power surreptitiously dumped one hundred pounds of live crawfish into the bathtub, tails snapping like canastas, and turned on the warm water, which almost instantly turned deep brown, the color of Louisiana bayou mud. He would drain and refill the tub as many times as it took for that water to turn at least as clear as Louisiana bayou water. The purging of the crawfish was a crucial, almost ceremonial prelude to a successful crawfish boil, something at which Power was masterful, yet humbly so. When Brody ran the boil, the mudbugs came out so hot your face went instantly numb, your tongue swelled, and your eyes watered, making the feast not so much enjoyable as it was survivable. Power's concoction of crawfish, cob corn, red potatoes, and seasoning had a tinge of the heat that did not displace the flavor. He then hung the *do not disturb* sign on the door, saving the motel staff from having a heart attack when they went to replace the towels and toilet paper, and went to work, because that day was going to be a busy one. His crews in Memphis had only a few hours to go into stand-down mode, to secure sites and equipment and job trailers to protect exposed installations before the weather hit. Because two things were blowing into town: *Dateline* and a blizzard the likes of which Memphis had not seen since 1998. It had been unseasonably cold all over the south so far that year, and I had climbed many towers with my Sony camcorder and interviewed shivering dogs from Atlanta to Nashville. But all the video of cowering dogs and blasting winds could not capture the kind of cold we were in. It just doesn't come across. You can't feel that cold any more than you feel that pain. How can one depict this? Just fourteen months earlier I was on Jimmy Tanner's stacking crew in southeastern Missouri. Between Halloween and Christmas of 2006, we had stacked several 240-foot self-support towers (complete with full complement of antennae and line) along the I-49 corridor between Jasper to the north and Joplin to the south. We had just one more tower to erect before Christmas, and that's when fifteen inches of snow and ice pummeled the region. Trees, heavy with ice, just gave up and snapped midway. The thousands upon thousands of downed trees were hauled off and placed in rest areas along I-44, which were then closed to travelers. Civil services were out to tens of thousands of people. The National Guard was rolling in to assist. The conditions were as bad as I had ever been in, but did we just call it and go home like sanity might suggest? Nope, dammit, we were gonna get this tower up before Christmas. The tower, laid out in twenty-foot sections, had frozen to the pasture ground, and we needed Lowells and boomers to nudge them out. Then we had to beat the ice and frozen mud from the structure with sledgehammers before we could fly our sectors with a crane boasting three hundred feet of stick. And just two days before Christmas, we topped it off. That day it was zero degrees on the ground with fifteen- to twenty-mile-per-hour winds. At 240 feet, the wind exceeded thirty-five miles per hour. The wind-chill calculator put "what the temperature feels like to your body"[71] at -27.4 degrees. Several coworkers and I were at 240 feet, and there is

---

71. THE NATIONAL WEATHER SERVICE wind chill indicator.

only one way to describe how cold it was: The next time the weather in your area drops to zero degrees, get in your car, head for the nearest open road, rev it up (*safely, please*) to thirty-five miles per hour, roll down the window, and stick your head outside for eight or nine hours. For the full effect, get someone else to drive and strap yourself to the car roof for eight or nine hours. Then take off your gloves so you can use your hands to work. Can you feel that? No, you can't. But Tim and I wanted to get as close to capturing that as we could. The impending Memphis blizzard was something we did not want to miss.

Tim and I were in Gadsden, Alabama, that morning, following Nike and Ludlow and the rest of her crew. Aliza had taken her camera and was exploring the home lives of some of our men up in Clarksville, Tennessee, and gathering backstory. But the Nike-and-Ludlow storyline was turning more into a soap opera than anything else. Ludlow was late for work. Ludlow didn't show up for work. Ludlow was making mistakes. Ludlow went AWOL. Ludlow was costing them money. The fact was Ludlow was using, but we did not know that at the time. By mid-morning Tim and I had had enough of this dog and pony show, and when we heard of the weather to the west, we cut bait and scrambled the five and a half hours over to Memphis. Unlike Nike and Ludlow, the Memphis contingent were known quantities to me, stalwart and proficient, and with Power running that market I knew it would be all about the work, sans petty histrionics. Also in market was Andre Broussard, the Ragin Cajun, a twenty-six-year tower veteran who specialized in municipal and government projects, fire, police, and 911. This was a pro who had special clearances to work even more sensitive sites pertaining to the military and homeland security. He was both colorful, as most Cajuns are, and eloquent in a manner unlike most of the other dogs. Andre had a natural ability to turn a phrase. I was hoping we could catch him on-site, but that hope dimmed as Tim and I pulled into the Red Roof Inn just before dusk and saw the parking lot packed with idling trucks and doors-open job trailers and crews milling about, stomping their feet against the cold. The snow had arrived just as we did, and the troops were packing it up for the duration. We found Power inside a job trailer he was modifying into a field kitchen, with tables and chairs and twenty-four-inch stainless steel paddles and basket hooks, with his big tripod propane burner and thirty-gallon pots. It was a 30 x 8–foot trailer, but he enlarged the area by erecting a 20 x 20–foot blue polyurethane tarp from the drop door into the parking lot. And though the wind howled and jostled the trailer to and fro, inside that cocoon it was toasty. The boil was on.

Power, as was his custom, gave me a big hug that wrinkled my spine while Tim introduced himself in his own disarming manner to the pack of dogs. I admired Tim's ability to put "subjects" at ease as much as Power's initiative in setting up this little haven in the storm.

"We could've used a rig like this in El Dorado," I said.

"Yeah, you right," said Power.

"Anybody left out there?" I said, meaning was there anybody still working.

"Nobody in their right mind," Power said.

That was demoralizing because Tim and I busted hump to get there. Then a dog chimed in and said, "Andre's got something going on." I shot a *What the fuck?* look at Power. He laughed, them big shoulders trembling.

"Like I *said*," he said, "*Nobody in their right mind . . .*"

Andre had a minor weatherproofing issue on a line he had recently installed for the police department, and he was en route to the site as we spoke. One look from Tim and I knew we were headed that way.

"Come back hungry," Power yelled to us over the roar of the big propane burner, which sounded like a jet engine inside that box trailer. The tower was only ten or twelve minutes away. Twenty minutes later, Andre and I were belted in at about eighty feet in the wind and the snow and the bitter cold, and he took off his gloves and went to work as I shivered and tried to hold my camera steady. Tim did not sit in the car with the motor running. He stayed out in the weather with *his* camera and stuck it out, which was commendable. That is what a real groundhog would do. You don't seek shelter when your top men cannot. I would estimate that by that time, I had shot over fifty or sixty hours of footage, but in the half hour we were on that tower, Andre captured the heart and soul of what I wanted this project to be in twenty-eight seconds. As the wind tore through the tower and the world beneath got frosted in snow nearing white-out conditions, Andre Broussard, his dark and cherubic face filled with purpose, laid it all out . . .

> *THIS is why I am here.*
> *THIS is what I am doing,*
> *THIS is why it is important.*

I had all I needed out of Andre and all I wanted from that tower, so I turned off my camera, and we hit the ground. As Andre and I squirmed out of our harnesses, he took his phone out, pointed it at Tim, and said, "This is what it's all about. This thing right here. For you to call Mama at the grocery store to get chocolate milk."[72] In that moment he was as much a hero to me as Terry Bradshaw or Paul Attanasio or Cady when he knocked out SeanDog.

Back at the Red Roof Inn, we *were* in a white out. Power was dumping out his second load of crawfish onto the newspaper-covered folding table, and the feast was on as a DeWalt battery charger/boom box blasted Led Zeppelin and Aerosmith. Tim and I had already had a twelve-hour day, but he was tenacious in his duty. He talked to everybody. He shot everybody. I huddled close to the burner as my fingers and toes tingled back into circulation and thought to myself, *This was a good day. We might really have something here.* Those feelings of camaraderie that had so eroded by the time I left Ramsey resurged in

---

72. See www.nbcnews.com/video/dateline/25738683/#25738665.

me. My favorite dogs in a motel parking lot having a crawfish boil in a veritable blizzard and laughing and bragging and kidding each other and hot food and cold beer and warm liquor and no work tomorrow and all was right in the world.

Later that night in my room, Tim informed me that he needed to head to L.A. to follow up on his other projects but also to sift through our efforts thus far and "see where the hell we are at with all this." Aliza would be returning to the offices at 30 Rock to file and categorize and edit down all that footage, and I was off the clock. I was to keep him in the loop, and if I thought something going on in the field had merit, I should give him a shout. He told me that he didn't need to hold my hand and that I could go out unescorted, pending his approval. That is exactly what happened.

A tornado took down a tower just outside Jackson, and I captured the wreckage and the emergency stack of a new tower. I visited sites where there had been fatalities just to get some B-roll of the terrain. I did sit-downs with Power and Sarge and anyone else who would sit down with me. Between road trips, I'd FedEx my discs to Aliza and touch base with Brody to see who was doing what, where. That's when I heard that Bo and Gunn and Super Mario and Bo's daddy John were doing a decom of a three hundred–foot guyed-wire tower in Oklahoma. This I had to see. Because they were not going to rent a crane and disassemble this structure twenty feet at a time—they were going to drop it. In the world of towers, that was about the most fun a dog could possibly have.

Harmon County, Oklahoma, is the southwestern-most county in the state, and therein sits the town of Hollis, where Oklahoma stops and the Texas panhandle starts. One of Hollis's claims to fame was Terry Stafford, who wrote and recorded the country hit "Amarillo by Morning" in 1973, which was fitting because Hollis was about fifty miles closer to Amarillo than it was to Oklahoma City. Online travel sites will call it a "close-knit" community, and I presumed upon arrival that it's easy to be close-knit in a town that most people had moved out of years ago. It had one stop light, two convenience stores, a roadside motel, a couple of restaurants, a bank, one elementary school that serviced the entire county, churches, and a municipal airport just north of town. King Cotton, which once blanketed thousands of acres in the region, was a profitable memory, and the principal product now was livestock. Its six- or seven-block downtown stretched east to west and was lined with one- and two-story businesses with false facades and once-hopeful cornerstones declaring *this* was established in 1904 or 1898. And almost every one of those businesses was empty. For a block or two north and south of the main drag, the downtown blight bled into the rest of the town, giving it all the charm and spontaneity of a waffle iron. To say it was a ghost town would be exaggerating, but not by all that much. On my ride out across the bottom of the state a majestic outcropping of rock would appear every thirty miles or so, or a looming mesa that turned colors as the sun rose and fell, but this was far from John Ford's iconic western vistas. In most places it felt like I was traversing the very surface of the moon, a dirty yellow moon the color of burnt straw. There was a beauty in this as much as there was in the vastness and sparseness of the Flint Hills. It was also a bit depressing. But Bo and Gunn and Papa John Cooper and Super Mario were up ahead, and I knew that just by being their indomitable selves they would have colored their landscape to suit them. They did not disappoint.

The L-shaped one-story motel was built sometime in the '50s, but the furnishings spanned every decade since: a Formica table from the '50s, a green cloth-and-wood "clinger" from the '60s, a starving-artist portrait of a schooner from the '70s, and, yes, textile mats of dogs playing poker. Each room was a time capsule unto itself. Aside from two other tenants, and they were *permanent* tenants, my boys were the only guests. In true tower dog fashion they had taken over the parking lot and the sidewalks outside their rooms with pop-ups and coolers and grills and chairs. The job site was just a few hundred yards away in a twenty-acre patch of weeds and fine, dust-like soil bordered by more acres of weeds. This land had been played out long before we got there. Across the parking lot was a restaurant typical of Midwestern farming or ranching communities. Open at 5:00 A.M. and closed by 2:00 P.M. At first this was a problem for the crew, who had been there over a week, because they were putting in long hours and by the time they got back to the motel the only thing in town to eat would come off the shelf at the convenience store. Microwave burritos, Vienna sausages, and Doritos had long lost any appeal. But they had become local celebrities—*those crazy tower guys down at the motel.* By the time I got there, which was just before dark, and up until midnight, locals cruised through the lot just to say howdy. A few goat-ropers would settle down for a beer. Mexican women came by with plates of tamales and refried beans. Teenaged girls in tight jeans and tank tops (in weather that though not as cold as Memphis definitely required a jacket) swung in to bum beer and cigarettes. They sported cowboy hats and big belt buckles, just like Gunn and Bo. The dogs were the only game in town, and they had ingratiated themselves so much to the community that the restaurant staff graciously gave Bo an extra set of keys so they could get inside after hours and "cook up a little something for yourself." This they did. They'd flop a steak on the grill and raid the walk-in for potato salad and coleslaw, leave a note regarding the damage, pay too much for it, and over-tip to boot. Though whatever glory days Hollis, Oklahoma, may have once had were long faded, the survivors were as friendly and kind and giving and trusting as you could hope for. These were the things that made the job tolerable, these unexpected buoys of humanity and decency amid a sea of poverty and imminent decay. These are the people both resigned and content to stick it out here for good or bad and not make Amarillo by morning. Perhaps they were holding out for the arrival of the fracking and wind farms that had turned the impoverished area of Oklahoma just 150 miles north into a boomtown where there weren't enough motel rooms to house the influx of workers and not enough convenience stores or bars or restaurants to take in the cash. I saw a lot of good in the parking lot that first evening, and with that, of course, came the bad. Which was that one of the two tenants there was a forgotten and sickly old man living on welfare vouchers and the kindness of strangers, sad and alone, doomed perhaps to die there. The other tenant was dispensing drugs to people who stopped by just long enough to roll the truck window down and move on.

After a week of prep the tower was scheduled to fall the next day. I called Tim to give him an update. I described the town to him, just in passing, and mentioned the tiny airport north of town.

"Airport?" Tim said, his voice rising half an octave. "Go rent an airplane."

The following morning I woke up a little fuzzy with the residue of the eight-hour drive to Hollis and Bo's crappy Natty Lite and the six hours spent in the L-shaped parking lot. *Did Tim really say go rent an airplane?* Indeed, he did. I walked into Gray Ag Aerial Services and told the man there what I wanted to do and asked what would that cost me, and the man, who was the pilot (who looked just like you wanted a pilot to look—tall, handsome, sober, slightly graying hair), said, "Hell, son, just cover my fuel." Up we went in a sleek yellow-and-blue single-prop crop duster. Bo and Gunn were atop the tower as we circled and circled half a dozen times, but the ride was choppy and I couldn't focus the camera, so I just stuck it out the window in the general direction of what I wanted to shoot and hoped for the best. On the way back to the landing strip, the pilot crackled over my headphones and said, "You wanna take the stick?"

By the time I arrived at the job site, Bo and Gunn were back on the ground. I pulled into the corner of the field alongside a dozen cars and pickup trucks. Many of the townspeople had arrived for the big drop. They lugged in coolers and set up lawn chairs or dropped tailgates. There was a carnival atmosphere about it, and I almost expected someone to saunter by in a red-and-white-striped suit and straw hat yelling, *Popcorn he-yah! Peanuts he-yah!*

There are three ways to take down a tower: one piece at a time using a gin pole or a crane, pulling it to one side and letting it free fall where it may, or telescoping it upon itself, which was the most difficult to get right and the most rewarding because it took some real skill. Bo opted for the telescope method. They had already removed all the antennae and radios and cable and lighting and support systems. They had also loosened the bolts securing the tower to itself at strategic locations. All that remained was the steel structure and the guy wires, which ran from three anchor points on the ground to the tower itself at various heights. When he was ready, Bo cautioned the spectators about staying *right where y'all is at*. Gunn and Super Mario then each took an anchor and overtightened the guyed wires so that extra pressure was put on two of the three faces of the tower. Then Papa John Cooper fired up his gas-powered chop saw, grinned a maniacal grin, said, "Dear Lord, if I kill somebody today, don't let it be me," and cut into the guy wires at the third anchor point.

Sparks from the composite blade of the saw on the steel wire threw up a fountain of fiery yellow and red. The guy wires popped and whipped into the tower. The tower itself groaned and then hissed, and it made its own wind as thirty stories of steel accordioned upon itself with a clattering, banging, squealing, screeching crash to the earth, sending up a final yellowish ten-foot-tall *POOF* of tired dirt.

And after a momentary hush—the crowd went *wild*.

It was a perfect fell, the debris field no more than sixty feet in diameter. I'd never seen bigger smiles on a crew in my life. Papa John threw up his face-shield and let out a shrill, south-Texas *whoop*. Super Mario did a little spastic fat-boy dance and slapped his thighs. Gunn threw off his hard hat, grabbed his cowboy hat, and did a deep bow to the audience. Bo just squinted and nodded his head. Sure, Bo and Gunn probably partied a little too much, and they had gotten me kicked out of motels and involved in fist fights I otherwise would have certainly avoided—but, man, they could sure drop a tower.

Many of the spectators stopped by the motel late that afternoon, sharing photos and videos they had taken of the big drop. I packed up, leaving my little room with the tile mosaic wall hangings of Dalmatians and kittens and the spring-loaded swinging-arm desk lamp, and I stood about in the lot for a minute. The crew would be staying behind because after they cut up and hauled off the crumpled tower to the nearest salvage yard, they'd be erecting a brand-new three hundred–foot tower in its place. And for the first time since I had been on towers, the scope of what we were doing hit me. Hollis, Oklahoma, which most of the country rightly or wrongly would consider to exist in the middle of nowhere, as well as thousands of towns across the nation even smaller and more desolate, were getting new and/or updated towers. They had *arrived* in a fashion, having the two things that seemingly permeated every nook and cranny of the forty-eight states I had worked in. They had cell phone service, and they had crack—divergent addictions, which outside of the insurance industry might just be the two biggest money-makers of that decade.

The sun was falling behind Amarillo as I nosed the Explorer eastbound on Highway 62. The outcroppings and mesas shimmered and morphed in the dying light. I didn't know it was the last time I would see Bo and Gunn and Super Mario and Papa John. Our paths would never cross again. But I did know I was out of stories to collect, that after four months in the field Tim and Aliza and I had pretty much gotten all we were going to get. Tim confided in me that he wasn't sure what we had would make a compelling *Dateline* hour, and that there was a good chance the whole shebang might be crashed indefinitely. Tim always laid his cards on the table, and as much as I wanted to, I couldn't disagree with him. This ride could very well be over. I resigned myself to that sad eventuality as I motored north through the Flint Hills toward home. It was a moonless night, black as a bull's heart, and with the exception of the sporadic blinking of tower beacons, the contours and depth of my beloved hills had turned to nothingness.

MAY 17, 2016

A Montgomery County, Maryland, police official has informed Wireless Estimator that the deceased worker who fell off of a tower in Damascus, Maryland, was 25-year-old Daniel Patrick Harrison of Indianapolis, Indiana. The public information officer said that Maryland Occupational Safety and Health (MOSH) was investigating the fatality.

Harrison, employed by Marali Telecomm Systems, Inc., had been working on an AT&T project being managed by turfing vendor NEXIUS.

In addition, the still-unnamed worker who was electrocuted in North Carolina was an employee of RH Construction, according to information obtained by Wireless Estimator. That incident is being investigated by North Carolina Occupational Safety and Health (NCOSH). A spokesperson said they could not provide any additional information until their investigation was completed. North Carolina has its own Communication Tower Standards as does Michigan.

The two fatalities raise the industry count to three in 2016.[73]

---

73. *Wireless Estimator,* "Two Tower Technicians Are Killed in Two Separate Incidents," May 17, 2016. wirelessestimator.com/articles/2016/two-tower-technicians-are-killed-in-two-separate-incidents/.

CHAPTER SEVENTEEN MISSISSIPPI
IT'S RAININ' MEN

At the end of March 2008, I sat at my perfectly appointed *just-so* home-office desk and stared at the Easter Bunny screensaver and rearranged my sticky notes and felt-tips and caddies because the little that was actually in my hands regarding the *Dateline* special was effectively now out of my hands. I was on an unwelcome hiatus. I tilled the garden, treated the deck with Woodlife, hung the bird feeders, readied the riding mower for spring, rearranged the furniture in the entire house, and drove Meg crazy because I was a man without a mission. The daddy time with my fourteen-month-old boy was irreplaceable, but other than that, I spent my days in what Thoreau had so perfectly pegged as "quiet desperation." I daily postponed, yet daily prepared, to call Jungle Boy and say, *Delaniac is not a writer yet again.*

Then on April 12, Charles Wade "Chuck" Lupton, thirty-four, veteran of the United States Army, recently engaged father of two, fell 150 feet to the gravel-and-concrete compound beneath a monopole in Wake Forest, North Carolina, killing him, hopefully, instantly.

Though catastrophic to his family, friends, and coworkers, Lupton's death was not at all unusual. It was expected. Statistically, his demise was behind the actuarial curve. In accordance with all records kept on tower-climber fatalities in the previous five years, Charles Lupton should have been the fourth tragedy of 2008, not the first. As Brody has said, "being number one is a distinction I can do without." Nonetheless, 2008 was shaping up to be the *safest* year for climbers since *forever*.

And then, as if some drastic, cosmic miscalculation residing within the laws of statistics and probabilities could not suffer to go uncorrected, men began to die at a pace unseen in the cellular-communications industry before or since.

Moorcroft, Wyoming, April 14: *A tower technician fell to his death in Moorcroft, WY, according to a Moorcroft Police Department administrator and Crook County Sheriff Steve J. Stahla. The worker was employed by Cornerstone Tower of Grand Island, NE. Officials of the Wyoming Occupational Health and Safety Organization were at a statewide conference this week in Casper and were unavailable for comment.*[74]

San Antonio, Texas, April 14: *James Friesenhaun, thirty-eight, was killed yesterday when he fell 225 feet from a guyed tower located near Northwest Military near Camp Bullis, in San Antonio, TX, according to Sergeant Ted Prosser of the San Antonio Police Department. The worker reportedly was loosening bolts on the steel that he was attached to when he fell. Two other technicians working on the tower said that they saw their coworker "sort of lean back a little bit, and apparently, after the last bolt that he loosened, he just fell down 225 feet," according to Sergeant Prosser.*[75]

74. OSHA and *Wireless Estimator*, "Fifth Fatality Has Industry Troubled About Sudden Rash of Climber Deaths Throughout the Nation," April 14, 2008. wirelessestimator.com/content/articles/?pagename=Tower%20Technician%20Deaths%202009.
75. OSHA and *Wireless Estimator*, "Second Tower Climber's Death in Past Three Days Mars Four Fatality-free Months," April 14, 2008. wirelessestimator.com/content/articles/?pagename=Tower%20Technician%20Deaths%202009.

Frisco, North Carolina, April 17: *A representative of the North Carolina Department of Labor said that William Edward Bernard, Jr. (Employee #1) of Chesapeake, VA, fell while working on a tower in Frisco, NC, last Thursday, and that additional details would not be provided until their investigation is complete.*

Employee #1 fell from an eighty-six-foot-high monopole cell phone tower while installing new amplifiers. The employer had a policy of one hundred–foot tie-off, and Employee #1 was known for adhering to that policy. However, Employee #1 was not practicing one hundred–foot tie-off at the time of the accident. Employee #1 was wearing a full-body harness and had a front lanyard (to keep him near the work site) as well as two lanyards attached to the rear D ring (as safety backups). Employee #1 fell to a sloped portion of an ice bridge, striking the bridge and falling onto a chain link fence, and then falling to the ground. Employee #1 was pronounced dead at the site.[76]

Natchez, Mississippi, April 23: *Mark F. Haynes, of Griffin, GA, died after falling approximately one hundred feet, according to Adams County Coroner James Lee. He said Haynes's death was caused by extensive head and chest injuries. Haynes was employed by Overland Contracting Inc., a Black & Veatch Company, and was reportedly hanging boom gates on a Cell South tower when the accident occurred. The tower tech was installing stacker blocks for coax cable within the interior of a cell tower between the eighty- and one hundred–foot level of the tower. As he was working his way down, the crew leader who was at a higher elevation noticed that he was not placing the coax into the blocks. The crew leader shouted to him*

---

76. OSHA and *Wireless Estimator*, "Fifth Fatality Has Industry Troubled About Sudden Rash of Climber Deaths Throughout the Nation," April 17, 2008. wirelessestimator.com/content/articles/?pagename=Tower%20Technician%20Deaths%202009.

to come back up and install the coax into the blocks as he
proceeded down. While the tech climbed back up, the crew
leader continued with his task. Several minutes passed by,
and all of a sudden the crew leader noticed the coax moving
erratically, and he looked down and saw his coworker fall.
The crew leader tried to grab on to the coax, but could not.
The tech was killed in the fall. From the investigation, the
tech may have been at the one hundred–foot level, preparing
to place the coax into the first block when he fell.[77]

Five deaths in eleven days.[78] Merciless gravity. The laws of statistics and probabilities were reestablishing the status quo and bitch-slapping us with a vengeance. And it was far from over. As expected, the phones between myself and Brody and Tim and Aliza were abuzz with the mindboggling developments of those eleven days. Any talk of permanently shelving the *Dateline* hour got up and left. The death I had so glibly promised the executives in Los Angeles seven months prior had arrived in spades. This was enough dreadfulness to (for the first time) make national media outlets cock their heads and say *What the fuck is happening out there?* as mainstream publications began to follow the trail of tears. A direct corollary of that was that the carriers began issuing the standard statements, such as . . .

AT&T Reemphasizes Zero-Accident Expectation
In an effort to reemphasize the importance of good safety
practices and AT&T's zero-accident expectation, AT&T
said that they were instituting a construction stand-down
to give each of their construction suppliers an opportunity
to reinforce appropriate safety practices with each of its
employees and subcontractors.

---

77. Ibid.
78. This was one fatality every 2.2 days. The only comparable surge in tower deaths that comes close is the period between July 8, 2013, and August 13, 2013, when six workers died, one every 5.8 days.

> *Their notice to contractors read, "AT&T therefore requires you to hold, at a minimum, a half-day safety refresher training course this week with all of your construction employees and subcontractors providing services for AT&T. Upon completion of the safety refresher training this week, AT&T expects that you will reinforce this training with additional random safety checks at the construction sites to ensure that appropriate safety measures are being used."*[79]

One would think that our *Dateline* contingent would have mobilized instantly and headed to Wake Forest, Moorcroft, San Antonio, Frisco, and Natchez, but we never even entertained that thought. Not once. These were not faceless names cataloged in the archived pages of *Wireless Estimator* and OSHA, as men like Jon McWilliams and Kevin Keeling had already become. These were lives and families now torn asunder for the most inexplicable and indefensible of reasons: 3G. From the Top of the Rock all the way down to me, it was unanimously decided to leave these poor people in peace, to respect these fresh wounds out of common simple human decency. That is exactly what we did.

But that is not to say the circumstances did not energize the project. Of course they did. Though we were not going to directly involve ourselves in the recent deaths, we knew they would be documented at some point during our program. They had to be. This was hard news, Tim's kind of news. It did not change the content we had accumulated, but it did amplify and magnify that content. Because now footage of Ludlow arriving late to a site with some bullshit excuse and Nike being all pissed off, and footage of Danny playing drums at the folk mass at his church, and footage of Ginger Jesus hanging benignly in his harness three hundred feet above Diddlypiss, Nowhere, carried with it the sobering import of recent events—that we all were what we always knew we were: one split second away from being bitch-slapped by statistics.

---

79. Ibid.

So did *Dateline* and I take the high road in not pulling some paparazzi bullshit at the expense of the newly and frequently fallen? We did. But that did not change the landscape for the end game; it only solidified it. It didn't even matter now how well the *Dateline* special was received because we had the one thing that hadn't been done in the collective history of work-reality programming . . . *the truck crashed through the ice*. Shit, a fleet of trucks crashed through the ice. Our human decency and respect for the dead was a conscious and altruistic stance, but we all knew there was a self-serving silver lining to it all. Though this was never outwardly expressed, you could see the relief and the excitement in the eyes of all involved. Behind the gasps of disbelief and the *"oh my God, you're kidding me"* and the eyes-down, head-shaking *tsk-tsk-tsking* was the knowledge that the blood-equity in a TV series about the real deadliest job in America where people actually died just blew through the roof at 30 Rock. I knew this to be true because it was true for me. I was giddy with the prospects, the *ka-ching* factor flittering in my brain like ticker tape. Though my shock at the frequency of the deaths was genuine enough, my sorrow and empathy were less than they should have been. I was one of them, and yet, I was not. This rash of fatalities not only solidified NBC's dedication to completing the special, but practically guaranteed a bidding war when Peacock Productions put it on the auction block. I knew this would happen, but I never imagined it would happen in such a remarkable fashion. And I had to justify this to myself. I was one of them, but I wasn't—because I was trained and followed that training. I was one of them, yes, but why should I care? These five men killed themselves, as the OSHA reports would irrefutably bear out. That is a fact. I did not know them. I did not kill them. They were not "one of us." And when yet another tower worker died twenty-three days later in Haubstadt, Indiana, it would have just been the icing on an already very lucrative cake had it not been for one fact.

He *was* one of us.

MAY 24, 2016

Kentucky State Police have identified a technician that fell from a cell tower in Morgan County as 19-year-old Tyler Comer of Charleston, West Virginia.

In a statement, authorities said the 19-year-old was working on a 105-foot cell tower when he reached for his tools and fell at 7:05 P.M. yesterday evening on a structure off of Hogg Branch Road.

The incident remains under investigation by Kentucky State Police.

According to a news report from LEX 18, the Morgan County coroner said that Comer was coming down from a cell tower when he fell 100 to 150 feet and that he was contracted through Appalachian Wireless.[80]

---

80. *Wireless Estimator,* "Kentucky Fall Claims the Life of a West Virginia Tower Technician," May 24, 2016. wirelessestimator.com/articles/2016/kentucky-fall-claims-the-life-of-a-west-virginia-tower-technician/.

CHAPTER EIGHTEEN MISSISSIPPI
BAREFOOT IN THE PARKING LOT

In death we made him famous.
Proud soldier dead at Carthage, lying honored on his shield.
Buonarotti, back to scaffold, eyes upon the chapel ceiling.
On a crowded lane in Padua, Mercutio, still joking with
his friends.
With ease we create Gods out of the dead.

The first funeral I ever attended in my life was that of my own father. I was nineteen. He was thirty-nine. He was a good man. He was a decent man and an above-average human being. But he was not Abraham Lincoln. Had my father lived, he would have continued to be a decent man and an above-average human being, but I doubt he would have cured lupus or built a better mousetrap. I wrote the above about my dad years ago, never knowing at the time how aptly it would apply to Triple J. He accomplished in death what he would never attain in life: He became a folk hero to what he would describe as his "peeps." And his peeps lionized him to the point of absurdity.

As in "To an Athlete Dying Young" by A. E. Housman, a poet even Richard Wilbur could not accuse of being oblique, to die in midstride not only dilutes and eventually absolves the indiscretions of your past but often assumes the forfeiture of a stellar future, a future packed with growth and accomplishment, a presumptive future as comforting to the survivors as it would have been unlikely.

Until I arrived at Triple J's funeral services in DeKalb County, Alabama, I thought myself not only to be well-versed but a bit of an expert on the topic of backwater and forgotten Americana. But I was wrong. I just assumed I knew about these places and these people because I had been here and there and I had seen this and that, but my knowledge as to how these people lived their daily lives was perfunctory at best. My contention that every little berg in America had at least one claim to fame was also shot to hell. Besides living, working, and dying, nothing notable ever seemed to have occurred around this part of DeKalb County. This place was not forgotten. It had never been known. But less than an hour to the northwest lay the town of Scottsboro, and that name rang a few bells.

Nike Rawlings, in a suitable black dress, and I, in plain black slacks and jacket, arrived to find that compared to Sarge Schmidt and Brody Kilfoyle, we were seriously underdressed. Both men wore impeccably tailored ensembles. Sarge in navy and Brody in gray. I had never seen Sarge in a suit, and it looked as if it would squeeze him to death, a stump-necked turtle in an overtight shell. My embarrassment over not wearing a tie was soon dispelled as I looked around the parking lot where perhaps twenty of Triple J's family and friends had gathered. It was then I realized not only did I not know shit about certain people from Alabama, but also that being in that parking lot I had might as well have been standing on a different planet. Aside from the representatives from K.M.C.A. and Triple J's uncle, Ricky Boots, the only person there dressed for the occasion was the funeral director. The other men wore jeans and a few polos but mostly T-shirts, and the women wore jeans and midriffs with the occasional thigh-high skirt and tank top. Footwear consisted of sneakers and cowboy boots. Many of the women didn't wear shoes at all. They had kicked them off at the entrance to the funeral home when they stepped outside to smoke or grab a beer or wine cooler from the back of one of the pickup trucks. I stood not in judgment of these people but in astonishment. Nike, herself from just an hour or so southwest of there, must have sensed my bewilderment because she took me by the elbow, leaned in, and whispered, "There's parts of Alabama just don't give a damn. They live their own stereotypes. The rest of us just shrug."

It was not Triple J the majority of mourners did not "give a damn" about. It was *us*. It was me. They were not putting on airs for no one, nowhere, no how. *This is us. This is who we are. Take it or leave it, we couldn't care less.* This really came home when later I saw Triple J in his open casket dressed in his bluest jeans and his favorite NASCAR T-shirt. He, too, was alien to me because though I knew him and had worked with him, I did not recognize him at all. He was waxen and void of any illusion he was once alive. Situated on a small tripod near his coffin was a plaque with a flowery sentiment referencing brothers in high steel conquering the sky, this overlaid atop a heroic photo of Triple J straddling a crow's nest, triumphant in the setting sun. That fallen-warrior sentiment was not comforting to me. It was unworldly, in a way, because the intended grandeur of it diluted the simple facts of what really happened to Triple J, Employee #1:

On May 16, 2008, Employee #1 was on top of a monopole telecom tower taking pre-job photographs. Employee #1 also rigged up the load line using an open hook and a simple block and tackle. After finishing these two tasks, Employee #1 began to rappel down the load line with a Fisk Descender attached to the front D ring of his full-body harness. As he descended, the open hook rolled out of the large carabiner

attached to the top of the platform. Employee #1 fell
approximately 150 feet to his death. The open hook, block
and tackle, and rope landed right beside him. The large
carabiner was still hooked to the top of the tower. The open
hook was missing the safety latch. The block and tackle
was rated at 1,250 pounds and was intended for lifting
equipment, not for life safety purposes. Employee #1 was
not using an independent lifeline when he descended via the
load line.[81]

Brody and Sarge and Nike and I stood quietly in the parking lot with everybody else, our hands folded respectfully before us, exchanging pop-eyed glances at each other, saying nothing but knowing what we were not saying: that Triple J had broken more rules of Tower Engagement 101 than we thought possible. He didn't do one thing wrong; he did *everything* wrong. He should never have been rappelling, and his foreman should have forbid that. His crewmates should have forbid that. He should never have used that damaged block[82] because they had perfectly serviceable blocks in their job trailer. Even when we did do a controlled descent, we *always* had a second safety line attached with a rope grabber just in case the first line failed, which in this case it did. He never pulled any of this crap in New Jersey or down in Philadelphia. There was no earthly reason for him to do what he had done. None of us could fathom this fatal disregard for the basics. Why on that day he chose to go renegade was something we would never know. It was Sarge who put it plainly when he said, "The fucking stupid son of a bitch killed himself. Simple as that." Triple J was wearing his belt. Triple J broke at least four (some say six) rules, and the enforcement of any

81. OSHA Inspection #311847461. www.osha.gov/pls/imis/establishment.inspection_detail?id=311847461'.
82. According to Gray Swain, safety manager for K.M.C.A., the block in question had been marked with a red X, meaning *REMOVE FROM SERVICE*. On their next stop by the main shop, that equipment would have been documented and then destroyed, as was and still is company policy.

one of them would have saved his life. "He had six ways not to kill himself and still managed to get it done," said Brody.

As harsh as those words might read, there was no malice in their tone or pronouncement. It was more a doleful explanation of the unexplainable. Both he and Sarge had been operating in some type of shock since the accident because after over thirty accumulative years in this business, they found themselves suddenly walking through their recurring and ultimate nightmare—they lost a man. They were innovators and problem-solvers and had no experience whatsoever in being victims of circumstances out of their control. Though they accepted the full bore of the responsibility for this, I don't think either man could rationalize accepting the blame. And, after a tower death, the first rule of business, after the lawyers on any side leap into action and scream for gag orders, is to assign that blame.

Do we blame the caller, the cell-service subscriber? Do we blame *you*? You were not there. Do we blame your carrier? They were not there. Do we blame the general contactor? They were nowhere near the place. Do we blame K.M.C.A.? The crew on-site was a sub crew with Sunburst Tower Systems. Do we blame that crew? The crew didn't kill Triple J.

He did.

And, amazingly, Triple J's last hurrah was not his first. "You'd think he'd'a learned after that first fall almost killed him," one of the mourners told Brody.

Apparently, while working some construction job prior to joining us, Triple J had spent several weeks in an ICU because he fell off a roof after refusing to wear his harness. He neglected to mention that when he applied to K.M.C.A. I have never seen Brody's eyes get bigger in sheer astonishment, and there was fury pulsing through his crow's feet.

I thought immediately of my phone call with Brody about Frogger: *"Is that goddamned waterhead alive because if he is I am going to come up there and kill him myself."*

That knowledge would have saved not only Brody from hiring Triple J in the first place, but most likely Triple J's own life.

Out of the 130 tower-related deaths documented in this book, thirty-eight, or 29 percent, have been attributed to some type of equipment or structural failure. Ninety-two, or 71 percent, have been attributed to simply not being tied off. But even those numbers are misleading because upon closer inspection, thirty-three of the thirty-eight equipment- or structural-failure fatalities were *still* the fault of the men doing the work at the time. Those thirty-three had toppled boom trucks and man-lifts, used underrated or illegal wire or rope for load and life lines, electrocuted themselves, brought entire towers down upon themselves, used improper anchorage points for their PPE, or overburdened safety hooks and carabiners. The other five fatalities were classified by OSHA as one due to cable-grab failure, two due to unexpected weather, and two where it is unknown why a load line just snapped (causing the deaths of Jerry Case and Kevin Keeling). Making that adjustment means that 96 percent of all these unfortunate workers did *exactly* what Sarge said Triple J did. And the blame fell right alongside Triple J.

And with that blame having been established, the next sidestep in this danse macabre was to enact the punishment.

The layers of insulation so imprinted in our industry are as much a force of nature as the gravity we challenge every hour of every working day. And when Triple J went down, he exemplified that fact: the shit falls down. The blame falls down. The i's are dotted and the t's are crossed in every contract signed by every participant in the chain of events from the conception to the design to the engineering to the assignment and to the construction of your cellular world. The deck is stacked. Every climber in America is playing a losing hand the minute he belts up and goes to work. The punishment, traditionally meted out in the form of official fines and an unofficial and unspoken (and patently illegal) loss of impending work, would land on the lowest level of the contractual chain, being Sunburst Tower Systems and K.M.C.A. The deepest pockets, the carriers, have the least liability, meaning zero liability. So who, then, in this multi-*multi*-billion-dollar juggernaut, looks after the tower dog? Who, besides the man on the tower hanging next to you, gives a one good goddamn?

Where is Crack Baby when you need him?

Ironically, the greatest and most outspoken advocacy of climber safety comes from the one entity most tower dogs would consider an enemy, the one royal pain in the ass most carriers and general contractors would gladly do without, and that is OSHA. Because OSHA is the only truly *independent* monitor of our existence. We do not appreciate or fully understand OSHA. They are the guys that make work harder in the attempt to make it safer—yes. They are the guys that add hours and hours to a normally profitable work-day, making it a losing or marginally profitable work day—yes. If we followed every rule OSHA has enacted into law, we wouldn't get a thing done in a timely, let alone profitable, manner—yes.

They are the best friend a climber has—yes.

Yet that climber has been indoctrinated into thinking they are somehow making his job more difficult. They are not. They are doing the best they can with what they have, which is not much. Like the Atlanta Police in Crime Zone 3, OSHA is understaffed and underbudgeted. This is a "new" industry, allowed for decades to run unregulated by anyone but itself, policed by nobody but itself. And as annual death tolls rose, OSHA did up their presence in the field, but that presence is usually felt after the lethal fact. Even after their investigators identified and fined the most egregious of offenses, these levies are frequently litigated, plea-bargained down to the offenders paying as little as 10 percent of the original fine(s). OSHA lacks the prosecutorial teeth to, as yet, make a profound contribution to worker safety, but not for lack of trying.

In April of 2015, OSHA sent out a nationwide RFI[83] asking *us*,

---

83. Request for Information: [Docket No. OSHA-2014-0018] RIN 1218-AC90 Communication Tower Safety. www.regulations.gov/docket?rpp=100&so=DESC&sb=doc Id&po=0&D=OSHA-2014-0018.

the tower hands, the tower workers, thirty-eight specific questions regarding the very state of our existence. Outside of Tim Uehlinger, nobody had *ever* asked us for our opinion about anything, certainly not one general contractor or carrier. This was a watershed document. They did this because they truly wanted to know what was going on in the field. They also knew they could not possibly go out there and find out for themselves, and they also could not accept as fact the posture and boilerplate assertions of carriers and general contractors. OSHA is nobody's fool. They wanted to get the skinny not from the jockey but from the horse's mouth. Workers were encouraged to reply to this questionnaire by mail, fax, or electronically, and it was online that I perused many of the thirteen hundred responses. Considering the size of our workforce, this was an enormous sampling of data for OSHA to sift through, and this was only the online portion of their RFI. There were thousands more. And every response I read dealt with the same issues we have been exploring in this book: low pay, long hours, harsh working environments, those pesky layers of insulation, unstandardized rules and safety regulations, an unpredictable workforce, pressure to complete, voracious middle management, and on and on. It was estimated it would take about two years for OSHA to interpret their findings and respond accordingly with (hopefully) new legislation. So, again, and until then, who in this multi-*multi*-billion dollar industry has our backs? One would think we would have our own, non-governmental watchdog within ourselves. Ostensibly, we do.

The National Association of Tower Erectors (or NATE) is the acknowledged public face of the tower industry. They are dedicated, in the parlance of their mission statement

- *To pursue, formulate, and adhere to uniform standards of safety to ensure the continued well-being of tower personnel.*

- *To educate the general public, applicable government agencies, and clients on continued progress toward safer standards within the industry.*

- *To keep all members informed of issues relevant to the industry.*

- *To provide a unified voice for tower erection, service, and maintenance companies.*

- *To facilitate effective safety training for the industry.*[84]

NATE has over eight hundred members, but these are not individuals. Men do not become members of NATE. Climbers do not become members of NATE. Corporations, LLCs, Incs, and Ltds become members of NATE. Aside from the companies that actually perform the work in the field, every major carrier, every major contractor, every major manufacturer, and every major supplier in the cellular-communications industry is a dues-paying member of NATE. Why is 3M a paying member of NATE? They make tape. Fuck, we use a LOT of tape. Why is Flash Technology a paying member of NATE? They make light bulbs. Lord, we need those. Why is Better Metal a paying member of NATE? They fab steel, and we use steel daily. What NATE *is* is a trade organization, dedicated mostly to just that. NATE is like a Rotary Club, and rather than being an independent association for the representation of workers' concerns, they are more a conventioneering collection of tower-company owners and hundreds of equipment and material vendors who supply the hardware and the technology (from entire pre-fabricated towers to the light bulbs that twinkle upon them) needed to make the (currently) 4G world hum. Though purported

---

84. NATE, natex.org.

to represent the welfare of tower workers, they have done little in their twenty-plus years of existence that illustrates that at all. They have a bimonthly magazine, the *Tower Times*, and they have a website. They have a board of directors. They have an office in North Dakota staffed by seven full-time employees. They have the peculiarity of being the only organization whose full-time job is to try to make some sense of an industry reinventing itself constantly in the quest to maximize profit. As an agent designed to promote, proliferate, and present a responsible and professional image of an industry to the world, NATE has done an exceptional job. They are to be commended for that. In a field where technical advancements are being made daily and initiated in the field almost as quickly, NATE has their competent fingers on the deviating pulses of all things tower. What they do not have is any tangible impact on anything that goes on in this deadly game above ground level. They have no more teeth than OSHA. And they have far less brass. Because whereas OSHA, in their mandate to keep people from being dead, actively searches for ways to make the industry more responsible to the individual worker, NATE pretty much admits in their mission statement that their goal is

> *... to provide a unified voice for tower erection, service, and maintenance companies.*

NATE's unified voice is that of seven people in North Dakota and a collection of over eight hundred dues-paying tower-company owners and vendors who supply those owners. Trickle-down towernomics. That unified voice seldom, if ever, reaches the ears of the men in the field. I have polled over two hundred on-the-actual-job tower workers and asked this simple question:

Me: What do you think NATE would say about that?
Them: Nate? Nate who?

NATE is as alien to tower workers as I was at an Alabama down-home funeral. That is because NATE communicates not down to the rank and file, but up through the vendors and general contractors and ultimately to the carriers. NATE knows the money to be made at the expense of the bottom comes from the top, and the focus they proclaim to bestow on the dog is given to the leash. They can do this because they know that the typical tower worker is unabashedly unaware of the service they provide every day. They know that tower dogs will never comprehend that they do not need NATE but that NATE needs them. NATE is less an industry watchdog than it is a lapdog. Because most safety initiatives proposed by or through NATE do not generate safety. They generate the *marketing* of safety.

You don't have to build a better mousetrap in this business to make a lot of money. You just have to build a different one. For argument's sake, let's say you build a new capstan winch (cathead) mount. (Even though catheads have been largely unchanged since the Crusades, innovators are fiddling with them constantly; they'll make a new rope safety-grab attachment, a new tie-off bar, a new method of mounting the cathead to the tower or truck.) This is all well and good, and if it works, that's even better. But then let's say you get that modification approved by OSHA and ANSI and then lobby a general contractor or, even better, a *carrier* into choosing that new doo-dad as a preferred—or even better yet—*required* piece of equipment to be used on all their future installations. Now you're really in business. You can apply the same process to climbing harnesses or lanyards or shackles or carabiners or utility knives or descenders, etc. All these roads will eventually lead through NATE. It is they who will inform the industry of these new requirements. Whether or not these alleged improvements are truly safer or merely a variation on an already-working theme isn't the point. The point is that once your new capstan mount becomes the required tool for XYZ General Contractors, installation companies will have to buy enough of them to retrofit their existing inventory of catheads if they want to work on those sites. K.M.C.A. owns twenty such catheads. The carrier will not absorb that cost, nor will the middlemen. The installer will eat that and will not receive one dime in compensation for yet a greater bite out of his bottom line.

This is not supply and demand; it is *you* provide or be damned.

The ultimate demand is the muscle on steel and the damned is the same. The demand that the carriers and NATE and that *you* depend on are the men who never before woke up and said *I think I will climb a tower today*, and then do exactly that. And if on any one given day these 8,500 to 10,000 men said *fuck it* and went four-wheeling, your world as you know it would be a very inconvenient place. If they said *fuck it* for as long as a week, you'd be stockpiling bread and bottled water. If I have ever learned anything in my whole forgettable fucking life, it is that I have *never* learned anything in my whole forgettable fucking life worth repeating, except for this: 350 million Americans depend on .002 percent of that population to get you through the damn day.

Much like the circumstances of his demise, Triple J's last words would become a matter of hearsay, conjecture, and debate. Two crewmen on-site could not agree whether he said *I'm okay, AM I okay?* or *How do I look?* The crew leader was not available for comment. He had vanished. We still cannot find him. According to Liz Day of *ProPublica*, who copublished an article in conjunction with a PBS *Frontline* episode airing over four years after the fact, Triple J's last words were

"Bouncy. Bouncy."

I do not know the source of that testimony, but I was in the company of Triple J's remaining crew, sans leader, and his family within a few days of the incident, and I heard no such thing. Yet I would not put it past him. Because Triple J was that breed of tower dog who by his very nature teetered on the razor's edge of asset and liability. He was young and strong, yet needy. He was confident, yet inexperienced. He was malleable to a point, yet stubborn in his life-long sheltered ignorance. But most of all he was fearless. He was as fearless as the six other tower dogs who would follow him to the grave before the year was out.

During most of the funeral service for Triple J, he lay alone with his plaque and his bluest jeans and favorite T-shirt. His baby mama and his baby and two or three other small children milled about the parking lot with the rest of the mourners and Sarge and Brody and Nike and me. It had become animated and loud among us all, which was as it should have been. Because at that point, all we had left was the company of each other and tomorrow. Triple J had no more tomorrows. The minister came out of the building and invited us all in for a prayer before Triple J was loaded into the hearse and driven off. We all went inside, and that was the only point that day everybody was in one room. Coworkers and bosses and family and friends were together for the first and last time because our world never works that way outside of the adhesive of death—maybe a wedding here or there.

Brody and Sarge invited everybody to a local restaurant for dinner and drinks, but before we did that, we stood in that parking lot just a little bit longer. The women kicked off their shoes, and we emptied the coolers, and some of the awkwardness between us faded into the common familiarity of loss. Ricky Boots and I hugged each other and shook our heads, still not believing any of this. Out of the tower contingent, Ricky Boots was feeling it most because it was he who, in an honorable attempt to make his nephew's life better, brought Triple J into the fold. No matter how much we told him it wasn't his fault, to him it would always be his fault.

Nike was helping Triple J's woman and that assortment of kids load up into a pickup truck and a minivan, and that is when I saw the mother of Triple J's little boy rub her hand through the curly blond hair of that toddler attached to her leg and say, in the tiniest broken voice, ". . . but who's gonna teach him to pee?"

JUNE 9, 2016

A New Mexico amateur radio operator fell to his death off a tower located on Mount Lemmon in Pima County, Arizona, yesterday morning. He was doing volunteer work on a radio club's equipment when the incident occurred.

Milt Jensen, 73, of Virden, New Mexico, had climbed amateur radio towers over the years, including his eight-circle array he owned just south of Safford, Arizona, as well as other clubs' structures to assist fellow DXers. It is not known if Jensen climbed professionally for a living following his retirement from the electric power distribution industry after a forty-year career. His LinkedIn page indicates that he had been self-employed since 2014. He is listed in business directories as owning Milt's Electronics in Lordsburg, New Mexico; however, the company is not registered in the state. Pima County Sheriff's Department spokesman Deputy Ryan Inglett said the man was working as a volunteer to repair the amateur radio tower when he fell at about 11:45 A.M. Jensen was pronounced dead at the scene. According to Jensen's son, Jason, in a post to QRZ.com, his father left behind his wife, Rulene; seven children; thirty grandchildren; and four great-grandchildren.

"He loved to help others, especially in his chosen hobby, amateur radio," Jason said.[85]

---

85. Wireless Estimator, "Veteran tower climber, helping ham radio operators, succumbs after a fall in Arizona," June 9, 2016. wirelessestimator.com/articles/2016/new-mexico-ham-radio-operator-helping-others-succumbs-after-he-falls-from-a-tower/.

CHAPTER NINETEEN MISSISSIPPI
THEY CALL THE WIND PARIAH

After Triple J's funeral, I spent the next few weeks assisting Tim and Aliza in trying to make a cohesive and compelling forty-eight minutes of show out of the mishmash of material we had collected. Most of the production and postproduction work was done at NBC's studios in L.A., just above the set where they taped *The Tonight Show*. Actually, *assisting* might be an overstatement. I mainly peered over their shoulders as they went from editing bay to sound booth, dubbing voice-overs and narration as Tim honed his script and Aliza and a talented cadre of technicians pieced together this jigsaw puzzle one-tenth of a second at a time. These were very gifted people who spoke in a techno-abbreviated-initialized language I could barely understand. They knew this, but they were not condescending and tried their best to tutor me regarding the process as much as time would allow. They had done hundreds of these hours. This was my first, and mostly I was just in the way. So I'd wander the lot and haunt the commissary, sitting outside sipping late-day burnt coffee just a few feet from where I found my filthy lucky duck, handing out Marlboro Lights to all the Angelinos who don't smoke. Some days, after the taping of *The Tonight Show*, I'd sit on one of the soundstage loading docks and watch Jay Leno stand outside and talk to any-body, and I mean *anybody*, who wanted a word or a photograph or a signature. He did this every day, always dressed in jeans and that blue collared denim work shirt. Thirty, forty, fifty fans. He took the time to visit with every one of them. Sometimes during the taping of the show, the pages would escort the entire audience to an out-door stage where the musical guest would perform. Most of these groups I did not know, but I remember getting to hear Los Lonely Boys. Tim would check and double-check his facts and his sources, talking to Brody and to Wireless Estimator. There were still a few things we needed, or wanted, to do. Tim, realizing my aerial foot-age of the tower in Hollis, Oklahoma, was subpar, meaning crap, had arranged for a helicopter to capture Cady and Brody on a tower outside Nashville. They had also produced several sizzle reels for in-house promotion and broadcast-ready promos (five-, ten-, and fifteen-second bits) to run in the days before the special aired. As I watched the piece slowly come together, I realized that it wasn't at all what I had envisioned from the outset. It wasn't bad. Far from it. Tim had a punchy, newsroom style, and the footage was fantastic. It was just *different*. Too little Ragin Cajun, too little Mrs. McWil-liams, too little Power and Danny, and too much the foibles of Ludlow. The foibles of Ludlow petrified me because out of all the employee and sub-employees of K.M.C.A., *Dateline* could not have picked a more detrimental and improbable representative of who we really were. They might as well have picked SeanDog or Frogger. Though I came to this conclusion far too late, Ludlow was the worst example of target labor—the payday junkie. He'd work until he had enough money to fix on whatever his fix was, and then come back to work when he was broke with a pocketful of lame excuses. This isn't a scenario exclusive to towers. I've seen it everywhere I have ever worked—the theater, the supermarket, the pipe shop, manu-facturing, and construction. It is simply a sad but true component of the American labor force. Still, Ludlow had done nothing illegal on camera or even really objectionable. I cannot even say his fix was illegal because I have no idea what it was. And though sporadic and unpredictable, Ludlow was more than just a loudmouthed tower dog. By all accounts he was a loving father, and he could be lovable himself in an oddly captivating and irascible way, some hang-dog-puppy-dog combination of man and child. I liked Ludlow Flagg a lot. He was just a part of the piece we could have easily done without. What we couldn't do without, and this was something I had been pressing Tim for since we started back in January, was to talk to NATE. I wanted their inclusion in this not only as a courtesy but also because I felt their voice should be heard. It is one thing for a tower dog to tell you his opinion of the state of the industry at dog level, but I, and Tim as well, wanted to at least get some take on the bigger picture. A little research revealed that a founding member of NATE and also a member of the Board of Directors was Kevin Hayden, who ran his company, Hayden Tower Service, Inc., out of a complex on Highway 24 just north of the Kansas River in Topeka. He was right in my breadbasket. Tim wanted to conduct the interview but couldn't make the trip, so I arrived at Kevin's offices one Sunday morning and set up my camera and mics. Tim phoned in, interviewed Kevin, and there it was. They talked for about an hour and though noticeably guarded (his PR consultant/lawyer—I couldn't figure out which—was present), Kevin was forthright and candid. Much of what he said could have easily been inserted into the course of the *Dateline* special as counterpoint or corroboration or even objection to what we were showing. But that didn't happen for two reasons. First, Kevin Hayden, though thoughtful and articulate, was too guarded and almost robotic in his demeanor. It just didn't play. Also, and this is my own damn fault, the video I shot sucked. The lighting and sound were so poor that the little bit we did use had to be subtitled. So 99 percent of that interview ended up on the cutting room floor, as did 99 percent of all our footage and interviews, 99 percent of Andre and 99 percent of Mrs. McWilliams and 100 percent of the family of Kevin Keeling and Power and Ginger Jesus and I could go on and on and on. The cutting room floor is often the realm of justifications and explanations, and I am offering neither. But somewhere in the archives of NBC *Dateline* sits a rolling cart crammed with over one hundred hours of discs and transcripts that were not seen and will never be seen, some of which definitely should have been seen.

The NBC *Dateline* Tower Dog Special aired July 21, 2008. A Monday. It cut both ways, and deeply. I will not detail or dissect or praise or condemn the show because you can see it for yourself and make up your own mind.[86] Quite magnanimously, NBC had flown many of the principals to New York City to view the show at Robert De Niro's private screening room in NoHo. My brother and sister and their spouses came in from Long Island, and the Three Wide Men crossed the river to make the viewing. Afterward we went to some pretentious NoHo bar; it was all faux brick and mirrors and snotty service, a shithole that was chic by geographic association. But we still had a ball. Tuesday morning fell way too fast and way too hazy but also brought a sweet meringue swirl of vagaries: Did Sarge and Jimmy and Scotty really take an NBC-proffered limousine back to Ramsey? Did my older brother Bob actually pay me a compliment? Did my sister Connie actually grasp my hand and weep that Mom and Dad would be proud? Did Meagan and I actually tunnel into the lustrous sheets at our posh midtown hotel?

Hotel?! I was in a four-star hotel! Not a crack-whore-meth-infested dump. Not a shithole on the vacillating edge of cellular coverage. Nike was in a five-room suite with her mom and daughter! It was (with the exception of the night my son was born) the greatest night of my life. I had worked for this night in one way or another for over fourteen years, and to be there as it was actually happening was otherworldly. I was walking in my own movie with my own selected

---

86. Watch it here: www.nbcnews.com/video/dateline/25786251/#25786251.

soundtrack—Blue Öyster Cult. This was the culmination of my college education; my fiction, theater, and film background; my desire to become what I was never meant to be; my promise to half a dozen dead relatives, including my mom and dad, that I would succeed in something no Delaney had ever done—to break out of the blue- to dirty-white-collar quagmire we were predestined to inhabit.

The news from NBC that morning was nothing less than astonishing. Before I could rub the sand from my eyes, I was informed that even *before the show aired* NBC had received offers from no less than five major cable stations looking for work-reality programming just based on the promos Aliza and Tim had crafted in L.A.—and the phone was still ringing. The reality show was a slam-dunk *reality*, and Brody and I were to meet with Tim and David Corvo later that day to discuss terms. It was the greatest thing I had ever done, and the sweet meringue quilt I had wrapped around me lasted all of, oh, say, about nine fucking hours. Then the shit hit a very non-oscillating fan.

This happened at breakfast. We were noshing on bagels and English muffins at a bright Manhattan diner near the hotel, beaming from the news we had gotten from NBC. Our cells were ringing and ringing with kudos and congrats and attaboys from coworkers, family, and friends. Then Brody took a call, and I could tell by his face it was not congratulatory. He lowered his voice, stuck a finger in his ear, and walked outside. As he stood on the corner of Thirty-fourth and Seventh Avenue, his comptroller in Nashville informed him that not only had a 1.5-million-dollar contract just been cancelled by a utility company, but that some of his crews in the field had been literally kicked off their job sites. This is because although the general public and the network programmers loved the show, the industry got peevish. Actually, the industry went as apoplectic as Sarge did when he throttled White Chocolate. The unvarnished spotlight on the men that do their bidding had shone where they did not want it to shine, and they reacted swiftly and with no small amount of venom and hypocrisy. And absolute power. With NATE in the vanguard, the industry as a whole decried the Tower Dog Special as an aberration, an embarrassment, a stain on their immaculate cloak of pious upstandingness.

This was not an overreaction; it was a feeding frenzy. Even some of my own coworkers, with whom I had toiled and sweated for years, accused me of costing them work, of taking food from their families' mouths. Some of the men I had loved and who I thought had loved me, this band of brothers, treated me as if I had just drop-kicked their kitten.

On Jungle Boy's end, things were worse: calls to the office cussing and harassing Lanai and anyone else who answered the phones; threatening emails sent to Brody and Sarge, calling them unthinkable names (which in our world is pretty damn hard to do); and NATE, posturing and denying.

I have been to a NATE convention, and it was not unlike any other convention where trade is done by day and by night the bars surrounding the venue are packed to the rafters with alcohol-sodden birds of a feather. It is there I heard war stories that make many revelations in this book seem tame by comparison. To have them tell it, these men built your world, they free-climbed it all, carrying one wrench and wearing only a loincloth. And why not? These men *were* the original Tarzans of the industry. These guys did do all the crazy shit we are not allowed to do anymore. They have as much right to brag and exaggerate as the tower dogs in the field today. But beware the pot calling the kettle black.[87]

So, in short, we got pasted.

They filleted us, one-tenth of a second at a time.

Though almost all of the viewers outside the industry were appreciatively appalled by what they learned about the cell world

---

87. "Funeral services are scheduled Tuesday for a 21-year-old Topeka man who fell to his death last week from a tower in Missouri. Jacob Von Kopfman became the fifth tower technician to die this year from falling from a communications structure, according to wirelessestimator.com. Kopfman was employed by Hayden Tower Service Inc. and was working on a tower in Missouri when he fell to his death." *Topeka Capital-Journal*, August 15, 2011.

regarding the death rate, they were also supportive, moved, and even entertained. I could paraphrase the hundreds of responses from those viewers in five words: "Man, I had no idea!" And that is exactly the reaction I was hoping for. Over five million people watched that program, which took second it its time slot behind *CSI*. That was five million cell phone users who learned about the high price of one modern convenience. One simple basic message had been sent: *know what your phone really costs.* Still the industry was unforgiving, scrutinizing the program frame by frame, picking out what they perceived to be small procedural infractions and making mountains out of what on any given day in the field would be a happily sanctioned molehill. I pacified myself by realizing that the criticism coming from the industry came from the smallest part of the overall viewership. But that didn't mean I was not both enraged and hurt by it. It was David Corvo who put it in perspective for me when he said, "Listen, I've done about a thousand hours of these things, and *not one of them* hasn't been bitterly criticized by somebody, somewhere. It is just a part of what we do. If you want to stay sane, don't engage these people. Don't get down in the mud because you'll never get out. Concentrate on the future."

That perspective, though comforting, was still like the cold you can't feel. Oddly enough it was Brody, who had just taken a massive hit in his wallet, who simplified the whole mess by saying, "One thing for sure, Delaniac—nobody can say it wasn't true."

I did take Corvo's advice. Well, *half* of it. Because I engaged. I got on the half a dozen or so industry websites where fairly vicious opinions were being zealously expounded, and I directly addressed many of the most outspoken detractors with what I thought was a sincere and reasonable argument for why they should maybe take a step back and at least consider the show for what it was—an objective look at our world at large. It was no puff piece, but it certainly was not a condemnation of the industry at all. And, of course, Corvo was right. Even if I was trying to appease these people, and I was not, there would be no appeasing them. They had become the righteously offended, as shiftable as Gibraltar.

Corvo's suggestion about concentrating on the future was also correct. Though Brody had to return immediately to home base to put out fires ignited by the program, I stayed behind at 30 Rock as NBC tangoed with potential suitors who wanted to air the tower dog reality series on their networks. The project was passed from *Dateline* proper to NBC Peacock Productions. The division, established just a year earlier, produced programming such as news specials, documentary specials, and reality series. While some of its productions did air on their flagship network, it produced programs for competing broadcasters as well, such as A&E and Nat Geo and Discovery, among others. And that's exactly where we fit in. Though at this time we really had no idea which direction this show would take as opposed to that of the special, one thing was for sure—it would involve Nike, because every bidder wanted Nike. She was the sizzle, the sex behind the sell. Nike, Nike, Nike. That was fine with me. She did deserve this. She was compelling and smart and tough and a joy to work with. And she was courted as diligently as Brody and I. But I also knew that after the industry backlash, I would somehow have to sever Ludlow from the mix. He was as unpredictable on our trip to New York as he was in the field. Like Frogger, he figured NYC was a great place to score, and he disappeared for a few hours, returning with a $140 bag of marijuana-scented mixed greens.

In the end it would be the History Channel (who had initiated and refined and made the reality-work platform a real winner) who engaged Peacock Productions to produce the show for them. Between August and September, meetings were held and contracts signed, and it was agreed we would begin shooting *Tower Dogs: The Series* in late October as winter set in and the conditions would be harsh and, therefore, presumably dramatic. I was assigned to an entirely new staff of five directly from Peacock, and though their professional resumes were impressive, this subject was beyond them. Aliza (who thankfully was assigned to the project) and I spent weeks bringing them up to speed with the very basics of the tower world. Tim had other fish to fry, but I really had hoped he would be supervising the project. Frankly, I don't think he wanted anything to do with reality programming. He had gotten close to us; maybe too close. And I think his old-school nose for news could not justify his talents being spent on what really was designed to be sideshow entertainment.

Special cameras and microphones were purchased and drones were being tested and real tower hands, Daryn and Crack Baby, were being trained to use those cameras. Though Brody put the entire Peacock field contingent through the same training we used to certify real tower climbers, it was decided the only people to ever go up a tower would be *real* tower climbers. And in deference to industry concerns, certified tower safety instructors would accompany all Peacock field crews anytime and anywhere they were ever near a tower. We did not want to make any of the missteps in the industry minefield that would further alienate NATE or the carriers or the general contractors.

Years before, when I worked at John Grace, we were awarded the contract for plumbing and HVAC renovations at Rockefeller Center, and we had a field office right above one of the magnificent neon Radio City Music Hall signs. And though I spent some time in that office, I never really could say that I worked at 30 Rock. I worked *in* 30 Rock, much like the custodians and the maintenance personnel, but I wanted to, I dreamed of, someday being a part of the inner-working mystique of that art deco masterwork.

"Where do you work?"

"Me? I work at 30 Rock."

"*Really?*"

"Yep."

It was the only dream I ever had that came true. I had a special pass, access to the floors where tourists were not allowed. Not that I wasn't a tourist—I was the biggest tourist there. I walked the polished stone halls and visited the gift shop, knowing that soon there could be *Tower Dog* mugs and T-shirts stocked alongside the merchandise from other popular shows. I visited the promenade between the sunken plaza and the eastern entrance, above the golden doors where the deco relief of Wisdom and Knowledge bespoke that *Wisdom and knowledge shall be the stability of thy times*. They did amazing things in that promenade on a daily basis. One day it was filled with the gleaming red and chromed trucks of the NYFD, dispensing out fire awareness and smoke detectors and kids' firefighter helmets. The next day they erected a cranberry bog, an *actual cranberry bog*, with two feet of standing water and thousands of cranberries and the antique wood and iron machines used to harvest them. In just a few weeks, the sunken plaza restaurant would be removed to make room for the ice-skating rink. And soon the big tree would be lit, and I promised myself I would take my son there on that very day, just like I was taken there as a child. Before then, I would be working in the field, all over America, with a full production crew of pros. I'd be shuttling between those off-the-path places and 30 Rock and the L.A. studios, nodding at Jay Leno daily and getting paid for all of it. *I really worked at 30 Rock.*

Beat that with a stick.

But the thing that really brought it home to me, the thing that made me really feel once again part of the New York I had moved away from so many years ago, was one morning when I was coming back from the deli with my arms full of coffee and Danish for Tim and Aliza, and I suddenly had to sneeze. I froze in the crosswalk and let out one hell of an *achoo*. A bag lady pushing a fully loaded shopping cart with warped and spindly wheels stopped, slapped me on the shoulder, and said, "Cover ya' mouth, ya' bastard."

Home again. *Home.*

As it would happen, and as you may have suspected, I never got to work at 30 Rock. We never got the chance to alienate anybody. Six days after the entire Peacock staff (which, including the safety mavens and top shooters supplied by Brody, had reached eleven full-time people) landed in Nashville with excitement and energy and about a quarter-million dollars worth of brand-new equipment—six days after Meagan and I and my son had moved into Brody's guest house, planning for the long haul—the project was cancelled.

Access denied.

We could not get permission to shoot on any of the dozens of towers where we currently had work, though Brody had previously been assured we would. Brody and I instantly thought we knew what had happened, assuming that some complicity between the industry and NBC and History had resulted in an understanding that it would be in their best interest if the show did not go on. What we didn't know was the mechanizations behind that decision. That is true to this day.

Could the carriers and general contractors have cajoled or pressured GE, who owns NBC, not to proceed? Possibly. Could Peacock (having realized the idiosyncrasies of producing *Tower Dogs* made the task more complicated and far more expensive than they had anticipated) have crunched their numbers and said, *We might want to rethink this one*? Possibly. Could History, whose latest work-reality rollout, *Sand Hogs* (which debuted September 2008), did so poorly in the ratings it was cancelled after a handful of episodes, have glimpsed a diminishing curve in such programming and decided to get out before they got in too deep? Possibly. It could have been one of these reasons or all of them or a combination of some. But most likely it was none of them. This was no conspiracy of the robber barons holding sway. The powers that be did not descend upon us with the fury of corporate greed and ass-coverage. This was just and yet another layer of insulation. Liability. Why would any tower owner in America, whether privately owned or owned through a corporation, want to take a chance of something going wrong on one of their towers?

Akrasia . . . *Why would one?*

In any case, it didn't matter because nothing could alter the fact that it was over. Mr. Toad's Wild Ride had sputtered to an ignominious halt.

The day the bottom fell out also had the ghoulish distinction of being my birthday. Brody, in some inexplicable attempt to cheer me up, bought a cake and had it decorated with little plastic skulls and tombstones. He shrugged off the cancellation of our show before it was even shot as just another one of those things. He had a fall-back position. He owned a company. He may have even been a bit relieved. I, however, was shattered. I was no longer a member of show business. I was an interloper who crashed the party and had done remarkably well up until now. And when all the producers and the field producers and production assistants from Peacock flew back to New York with a quarter-million bucks of equipment, it was just to wait until another assignment crossed their desks. The only thing crossing my desk would be bills and the knowledge that you only get so many shots at success in the biz, and I had most likely blown my last one. No Marc Shmugers or Tony Bills could tweak this calamity. I could never again say that I worked at 30 Rock. I no longer worked anywhere.

NOVEMBER 2, 2016

A 44-year-old man (Steven Lamar Hill of Chandler, Oklahoma) is dead after falling hundreds of feet from an east-Tulsa cell phone tower. Tulsa police say the man was working on the tower and slipped and fell near Thirty-first and Memorial around 1:30 P.M. Police are investigating. Officials are unsure if the man was harnessed.

The owner of the tower, SBA Communications, has been inspected by OSHA ten times since 2000. They have not reported any previous fatalities. Police said the man worked for Michigan-based Mann's Tower Service.

They have been hit with at least seven violations in the past, including three listed as serious.

This is the sixth cell phone tower worker killed in the United States in 2016.

Tower workers are ten times more likely to die on the job than construction workers, according to the Department of Labor.[88]

---

88. Fox23 News, Tulsa.

CHAPTER TWENTY MISSISSIPPI
I AM THE GREATEST THAT EVER WAS

One hundred and forty-seven years before we drifted into Abilene, Kansas, John Wesley Hardin did the most famous thing he ever did. He shot a man for snoring. Some say through a wall. Some say between the eyes. Some say he knew the poor victim and did not like him. Some say it never happened. Again, like Triple J, accounts of that occurrence are as varied as they are unreliable. But this much is true: this was still cowtown Kansas, and we rode in just like the men the cellular industry had so vehemently described us as: noble, taciturn cowboys, riding the high stony lonesome, loping in from weeks on the open range, a vital component of American commerce. The cowboys bring you meat to eat, and we bring you cellular service. Two absolute necessities. We are here to hook you up.

Glory be.

That description is about as romanticized and inaccurate as the old West itself. But it is also, in the words of Hemingway, *pretty to think so*. What wasn't pretty was us as we moseyed on into town. We had been stacking a five hundred–foot guyed-wire tower about twenty-two miles northeast of I-70. The plains were high and cold.

Kansas is indeed many countries, and some of it is just butt-ugly. The soil was like sand, and the sagebrush hid rattlers and gopher holes. One of our crew had already broken his leg in three places in one of those holes and was airlifted to Topeka. It was a shit site on a shit day in a shit place, and all we wanted to do was get a room and a shower and some sleep.

But the ghosts of drovers past made damn sure that didn't happen, and before the night was over we were "rolling around in the blood and the mud and the beer." Johnny Cash and Hemingway: alpha males. We had not rode into Abilene to raise hell. Hell came to us that night in the form of our neighbors at yet another crap motel. They were not staying at this place; they were living there. They were addicts, disenfranchised, meth-soaked, and loopy inside their own conception of reality. And they fucked with one of our trucks, scratching with stringy, claw-like hands against the windows and doors, seeking a seam they could pry into, like crabs into a mollusk, most likely in search of a Garmin.

I tried to warn them off. I tried to dissuade them. I threatened them with calling the motel manager, at which prospect they just laughed, and then I threatened them with the cops. At this they also laughed. I was in Abilene's version of Crime Zone 3. When I yelled at the top of my voice that *if you don't stop messing with that truck I will kill you*, one of them slithered into the darkness. The other came right at me, brandishing a dried-up paint brush, which I assume he intended to be a weapon, and he proceeded to spit on me, swearing he had AIDS and that if he wasn't going to paint me to death he was going to infect me with his disease if I didn't back off. He was a horrifying figure. Teeth rotted and black. Emaciated and sunken-eyed. He was shirtless in the twenty-eight-degree weather and sweating bullets. He was scarred, and only short sporadic tufts of hair remained upon his dented head. And that head was like a shroud of dried, flaky skin on skull.

I had not hit another human being in anger since I was nineteen years old. But that night I dropped this motherfucker like a rock. I hit him hard, and I hit him more than once. He crumpled into a sniveling ball of hopelessness and began to weep. He crawled inside his room next door. Moments later, washing his filth from my bruised knuckles, I, too, began to cry because I was ashamed. I did not have to hit him. I had inveigled my way out of dozens of these situations before, but with my words, not my fists. I hit him because I wanted to. I hit him because life had become insufferable and I blamed him. I had disconnected as much from myself as I had from any sense of normalcy or decency that existed in the real world, the world in which I no longer belonged. The primer was a clouded memory. It had been three years since the tower dog show was cancelled before it was even shot, and I had fallen into and had been living in what I only can describe as self-rationalizing and indescribable despair. I was a goddamn wreck, yet a functioning working wreck, when I felt my fist break into his cheekbones, hearing them pippety-pop like bubble wrap.

And then the pounding started. The thin walls of my room began to shutter and the plaster began to crack. The wall hangings fell and the lights flickered. Muffled, unintelligible words emanated through the heating vents. Just as suddenly as it started, it stopped.

I took a long hot shower and put on some clean jeans and fresh socks and a warm hooded fleece sweatshirt. I peeked outside and saw that the man's door was still open. I could hear him whimpering. I went inside, and he was cowering in the bathroom, wrapped around the base of his vomit-pasted toilet.

"Please don't hit me," he said.

"I won't," I said. "I promise."

As much as I did not even want to touch this man, I grabbed him under the arms and dragged him to his bed. It was then I took a good look around the room. I saw four walls covered with smears and smatterings and elongated drippings of blood. His blood. He had been hurling himself against these walls, pounding his pain away. An ambulance arrived moments later, summoned by whom,

I do not know. One EMT, her face struggling between compassion and disgust, told me he did indeed have AIDS and he had been trying to kill himself on a regular basis. This was not her first visit.

"He just isn't any damn good at killing himself," she said, snapping the rubber gloves from her hands. "He just isn't any damn good at it at all."

To say I had changed in the three years since the *Tower Dogs* series had been squashed would be accurate. To say I had gone off the rails would be more accurate. Though I tried to be stoic and slough it off as just another professional rejection in a life filled with professional rejections (much like Brody had), the truth was I could not rise above that failure as I had the failures of the past. The failures of the past might be a few bucks here and there: a movie greenlighted but never made, a play published but sparsely performed, a short story published but never anthologized, a movie made but never widely released. They were all acceptable losses. It would take me over eight years since October of 2008 to consider the entire tower dog project, from inception to broadcast and beyond, twenty-four years of my life in total, as an acceptable loss. It was just too damn close to me to let go.

Hell, I am not sure I am over it now.

The day after I blew out the candles on that macabre cake my best friend festooned with trinkets of doom, Meagan and I and our baby left Nashville for the temporary security of our little leaning house in the eastern-Kansas hills. Then we split up. I took what little I had saved from field producing and borrowed a lot more and invested in several self-propelled business ventures, including a hot dog stand and a little tavern that also failed outstandingly. Soon, I knew my financial deliverance was once again all things tower. But I could not go back to Brody. Though I knew he would take me in with open arms and a wink of those blues and say, "This is where you ask me for two weeks per diem up front," what little pride I had left prevented me from even asking. So for those three years I ran with another company, the *we* that rode into Abilene that night. These men I worked with would make the *Dee*-troit City Cowboy, Ludlow Flagg, look like Pat Froggin' Boone.

If NATE thought that the *Dateline* portrayal of tower climbers was an inaccurate portrayal of their labor force, they should have seen these boys. Because these boys were the true majority of the men I have met in the field. These are the men that are the true makeup of the tower-labor world—the absolute fucking lunatics. If NATE wanted choirboys, they should have embraced the men we gave them in the *Dateline* special. The boys I ran with now were the dark end of the cowboy spectrum—the boys that tore up the saloon, abused the saloon girls, drank whiskey by the bottle with a cork in their mouths, and spit on the floor. I had seen a lot but nothing like this. If I told a motel owner our company name, I was shooed out the door with a broom and profanity. In Mississippi, at a camp-site, these boys were going through a jug handle of Crown Royal a day and hanging the purple-and-gold bags as witness to their alcoholic prowess all over the place. When the park ranger kicked them out, he said, "You are the worst campers *ever.*" In a lovely hamlet in northern Wisconsin, the kind of town with seven hundred inhabitants and six bars, these boys managed to run out on every tab they ever ran up, prompting me to make the rounds on a nightly basis and say to every bartender, "Okay—who owes you what?" And this was not the exception of tower crews. This was the real rule. When I laid hands on my neighbor at the crap motel in Abilene, these boys did not have my back. Power would have had my back. Ginger Jesus would have cut him off at the knees. Brody may have shot him. But when I looked over my shoulder for support when I was pummeling this poor bastard, and looking out of one eye for his friend, my new crew was too fucked-up themselves to care.

So I drove out that night, headed back east to a house without a woman or a child, thinking to myself, *If we are the cowboys of this technological age, who will need or want us when this boom is over?*

And it will be over. The train and the truck ultimately replaced the cattle drive, and we will be ultimately replaced by technologies not too far away, closer than we think, and closer to the ground. Solid ground, where any slap-dick can go to work. Some of us might end up permanent residents in some nameless motel in Hollis, Oklahoma, or Abilene, Kansas, or Ramsey, New Jersey.

I jogged off of I-70 into the brittle Kansas night, now sleeting with snow snakes across the blacktop, and I slid into the lower roads, the blue highways of William Least Heat-Moon, the small roads in a small world where towns lay miles apart and where people went to sleep at a decent hour and got up at a decent hour and just had a normal day. I drove slowly through the capillaries of the heartland, savoring every mile, free from that maniac crew but still worried about making a buck. I drove slowly because I had no place to go besides that empty house. I had never been lower in my life, and, go figure, it was a tower dog who pulled me out of it.

Because the Godfather called.

Salvation comes when you least expect it, perhaps prompting the word. Being *saved* should always be a surprise. Is salvation what you want or what you need? Is salvation an unexpected realization that shit ain't so damn bad after all, or is it that you have come to accept that shit *is* really that bad and you have just grown enough, or matured enough, or have been beaten down enough, to become accustomed to it? Is salvation just accepting the lot you have been dealt, and, being dealt that lot, to do your damnedest to overcome it?

You tell me.

My salvation was accepting failure. Not a tough nut to crack. I accepted I would never, ever, walk in my dream life and raise my son in that dream life of prosperity and the annual lighting of Christmas trees and the daily miracle of cranberry bogs and fire trucks at 30 Rock. But I could fall back on the destiny that was and had always been mine—my fearlessness of hard labor. And the Godfather and I went to work. And in that labor I was reborn.

Ludlow Flagg foreswore he was the greatest, the best that ever was. All dogs are that. I have never been at a motel parking lot debriefing where the question, or challenge, did not arise as to who was the best, the greatest tower dog of them all. It was, of course, whoever was talking at the time: Who got hit by lightning? Who rode out the harshest storm? Who was on a tower the longest? Who pulled three one hundred–hour weeks in a row? Who could climb the fastest? The highest? Who worked in the coldest environment or the hottest? Who had fallen and survived? Who got injured on-site and carried on? Who had whose back, and where did they have it?

I have been on a tower hit by lightning. I have sat with Brody at four hundred feet while daily thunderstorms rolled through us above Valdosta, Georgia, and we got lit up by lightning twice. The hair on our heads and arms stood up, and we just locked eyes to see if either one of us had died. I have done consecutive one hundred–hour weeks and countless ninety-hour weeks. I have been on a tower in Cincinnati where it was twenty degrees on the ground with thirty-five-mile-per-hour winds up top. I have been on a tower in the Texas Panhandle, when it was 110 degrees all day for five days. I have gone up a tower at 7:00 A.M. Wednesday and not come down until 3:30 A.M. Thursday. I had my head cracked wide open by a piece of fallen steel in Kingdom City, Missouri, and after patching up the gash with Super Glue and two-inch-wide electrical tape, I went back to work. I have no memory of that because I was knocked out senseless, but the God-father told me later it was "ballsy." I have fallen, and I have survived, because though I was careless, I was not reckless. I fell into my gear.

Am I the greatest?

No. Not for those reasons. Not by a long shot. That is just the job. The greatest are those who do it and wake up the next day and do it again and again and again. These men are as much heroes to me as Terry Bradshaw and Paul Attanasio ever were. Still, for forty-five and a half days in the summer and fall of 2011, I was greatest of them all, and this is why:

When the Godfather called me, he had just secured a contract with a transcontinental pipeline to inspect every tower running along every pipeline they had coursing through our lower forty-eight. And there were a lot of them. By law, every self-support tower in America has to be inspected every five years, and every guyed-wire tower has to be inspected every three years, and this pipeline had plenty of both. We hit the road, just me and the most knowledgeable man I had ever met in the business.

In the forty-five working days between July 10 and October 8, 2011, I climbed 147 towers, averaging 3.5 towers a day. One day in Oregon I climbed eight towers. I climbed an average of 406.5 feet per tower and 1,425 feet per day. I climbed over 59,383 feet, or 11.3 miles of steel. We rolled through twenty-five states, averaging 327 miles per day, over twenty-eight interstate highways, hundreds of state and local roads, and just as many goat paths. We rolled from Louisiana up to Michigan, from Louisville out to Dodge City, from Arkansas to Washington State. We passed through forty-five major cities, four of the five Great Lakes, thirty-two major rivers, and twenty-seven mountain ranges; we saw forty-six different types of indigenous wildlife; and we stayed in forty-one different motels.

It was done efficiently, professionally, and profitably, and it was the most satisfying work I have ever performed as a tower dog.

And for those forty-five and a half days I damn well was the greatest.

I was the greatest you goddamn ever saw.

ABOUT THE AUTHOR

Douglas Scott Delaney is an award-winning and produced playwright and screenwriter. He has done development work with Fox Searchlight and Columbia Pictures, among many others, and has done production work for both TV and film. His one produced film, *All Roads Lead Home*, won Best Picture at the Los Angeles International Family Film Festival in 2007. His short fiction has been published in *Kansas Quarterly*, *PRISM International*, and *Western Humanities Review*. His drama has been published by Samuel French, and his plays have been performed in theaters all over the United States. Delaney was born in Brooklyn, New York, and raised in Levittown, Long Island. He now lives in the Kansas Flint Hills, where he is "at least a quarter mile from anyone who could aggravate me, and vice versa." He has been a tower dog since 1997.